CHURCHILL TANK

1941–1956 (all models)

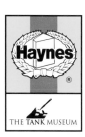

COVER CUTAWAY:

Churchill Mark III tank. *(Ian Moores)*

First published in August 2013

A catalogue record for this book is available from the British Library

ISBN 978 085733 232 5

Library of Congress control no. 2013934890

Published by Haynes Publishing,
Sparkford, Yeovil,
Somerset BA22 7JJ, UK.
Tel: 01963 442030 Fax: 01963 440001
Int. tel: +44 1963 442030 Int. fax: +44 1963 440001
E-mail: sales@haynes.co.uk
Website: www.haynes.co.uk

Haynes North America Inc.,
861 Lawrence Drive, Newbury Park,
California 91320, USA.

Printed in the USA by Odcombe Press LP,
1299 Bridgestone Parkway, La Vergne, TN 37086.

CHURCHILL TANK

1941–1956 (all models)

Owners' Workshop Manual

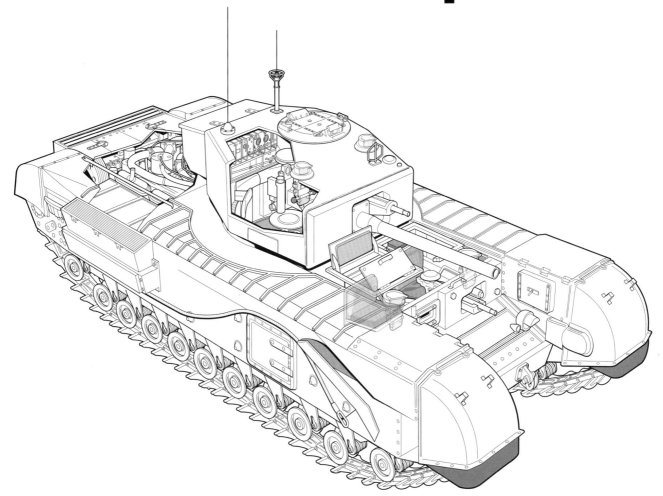

An insight into owning, operating and maintaining
Britain's Churchill tank during and after World War II

Nigel Montgomery

Contents

OPPOSITE **The Churchill Project's Mark IV is re-stowed before a demonstration run. The boxes on the track guard to the left of the photo are for Besa machine gun belts. Each wooden box contained two metal 'liners' with a belt of ammunition inside. The tools leaning on the other side are stowed on the engine decking.** (Author's collection)

Foreword

BELOW A Mark IV
or VI with 75mm gun
advances warily. Note
the extra track links
as unofficial armour,
including over the
driver's vision port.
There appears to be
an additional piece of
armour welded beside
the commander's
cupola to protect
him from sniper fire.
The driver's and front
gunner's periscopes
have been raised to
'see' over the track
guards. (Tank Museum).

It may seem strange to write a manual for the Churchill tank some 68 years after the last one rolled off the production line to find its way, eventually, to the Tank Museum in Bovington. However, interest in the Second World War thankfully remains undimmed and the story of these tanks matters because of the pivotal role that they played in Britain's contribution to that war.

This book is a grateful tribute to all those who served in Churchills as well as those who helped to keep them running and equipped.

It aims to provide useful and verified information for those who want to know more about the tanks, whether from a historical perspective or as modellers or wargamers. I hope that the book may also inspire others to seek out and restore some of the remaining Churchills, and perhaps encourage those who already have examples and are wondering whether to work on them.

This is indeed a manual, in that it explains how to restore and service the tanks. It is a guide to the tanks themselves, describing some of their history and fighting roles. It is based on the actual restoration of Churchill tanks by the Churchill Tank Project, which aims to promote and maintain knowledge of these vehicles and makes them available for hire for films and documentaries. The Project is producing a DVD of the restorations, with much information about the tanks using current and historical film. More information about the machines and, in due course, about the DVD, can be found on the Project's website at www.churchilltank.com.

I would like to acknowledge the help given by the Tank Museum at Bovington, in particular David Fletcher and Janice Tait, the Librarian, and the Curator David Willey. David Fletcher, aside from writing the Introduction, also provided much of the material on the Churchill AVRE and Flail in Chapter 2. My thanks also to Richard Morrell for giving me access to his family archive on the NA 75 (page 60) and to Tim Strickland for allowing me to reproduce his father's lecture on Churchill tactics (page 144).

While mistakes are mine alone, I would also like to thank my family for helping to proofread and comment on the contents.

Nigel Montgomery,
Cambridge, New Year's Eve, 2012.

A NOTE OF CAUTION FOR THOSE WORKING ON CHURCHILL TANK RESTORATION OR MAINTENANCE
The tanks contain hazardous materials including asbestos (not least in brake linings, clutch plate, heat shield on the engine and original cable insulation) and radioactive luminous paint (on the radios). Working on the tanks can be dangerous and proper advice and guidance should be sought before undertaking the tasks set out in this book. Regulations regarding ownership and use, as well as handling, storage and disposal of waste should be complied with.

Introduction

**David Fletcher,
Historian at the Tank Museum**

Born at the turbulent opening of the Second World War, the Churchill went on to become the best British tank of that period. In spite of all the wrangling and uncertainties that preceded its creation, the Churchill became a real success story for its time.

About 13 years ago, when I was writing a book on the Churchill tank, an American publisher friend of mine suggested that the title should be *Churchill, the British Tiger* or something like that. A bit tongue-in-cheek maybe but it was not such a bad suggestion. Both tanks had thick frontal armour, both suffered reliability problems at the start of their service and both confronted one another, in action, in Tunisia in 1943 and onward through the war. However, that is where the similarity ends. While, in terms of firepower, the Tiger outclassed the Churchill dramatically, the Churchill played a pivotal role in Britain's war effort, and thus victory for the Allies, while the Tiger, too few in number and always somewhat unreliable and fuel thirsty, did not stop the tide.

In any case I was not planning to use that title because I felt that the Churchill tank had a story of its own to tell and should do so standing on its own tracks so to speak. Counting on the word Tiger to boost sales

BELOW Former enemies. This Churchill Mark IV is exactly like the tank that knocked out this German Tiger in Tunisia in 1943 (see page 67). Tiger '131' is the part of the Tank Museum collection at Bovington and is the world's only Tiger I in running order.
(Author's collection)

would simply not do. The fact that Haynes Publishing are prepared to support a book devoted to this tank shows that the Churchill has achieved recognition in its own right, as it should, since it was arguably the most important British tank of the Second World War.

Its birth and even description was founded in some confusion, not least because British tank philosophy itself was far from clear. As the fear of imminent war grew, those in authority reverted to their earlier experience and started to worry about another Great War. All they could see was a new Western Front with even more formidable fortifications and as a result began to consider the creation of a new 'assault' tank, bigger and more heavily armoured than any before. The first of this intended breed of larger tanks, given the General Staff specification A20, was originally considered in September 1939. It embodied a mixture of ancient and modern concepts which, taken together, were deemed incompatible by the technical experts and in the end it had to be entirely redesigned. For example, the original specifications called for a turretless tank so that an undeleting beam could be deployed, but the experts were convinced that a turret was crucial and, indeed, by the time every one of the General Staff's requirements had been examined and rejected the tank was essentially an entirely new design, although it still retained the original General Staff designation. This set a pattern of compromise that tended to dog British tank design for many years.

Two prototypes of the redesigned tank were ordered from the Belfast shipbuilders Harland & Wolff, builders of the *Titanic* among other things, and these would be followed by a production order for 100 tanks – about enough to complete two battalions. In the event, though, only two were completed as running hulls, one of which sported a Matilda tank turret, for want of anything better. By January 1940 it was clear that Harland & Wolff were struggling to meet this commitment alongside all of their other work, not helped by the fact that the authorities were still unsure what weapon to recommend for the new A20 and where to mount it. To help, Vauxhall Motors of Luton was engaged to assist Harland & Wolff and to design a new engine, since this was another area in which difficulties were being experienced.

In May 1940 the two pilot models were nearly ready for trials and the second one, A20E2, had been shipped across to Luton. However, May 1940 was also the month of the German invasion of France and the Low Countries which, at a stroke, rendered the need for heavy assault tanks redundant, at least for the time being. The Germans had shown that aggressive use of tanks and other armoured vehicles, supported by motorised infantry and dive-bombers, could sweep across the ground, leaving fixed lines of fortifications isolated in their wake and spreading terror wherever they went. A new age had dawned, one that ironically had been forecast by the British Mechanised Force experiments of the late 1920s, which the War Office had elected to ignore and the Nazis eagerly absorbed. It was an age in which there was no place for the assault tank; the A20 project was cancelled.

Although the original A20 concept was now a dead duck, Vauxhall Motors still had the prototype for which they were developing an improved suspension and a new engine. It was obviously foolish to waste the development work that had already been done.

Against this background, the first meeting of the newly constituted Tank Board took place on 24 June 1940, and item one on the agenda required those attending to 'consider the specification for proposed new Tank A22' and they had three days in which to do it. All of this haste rather suggests forceful direction and a sense of urgency backed by someone in authority, probably the Prime Minister. Winston Churchill succeeded Neville Chamberlain to that post on 10 May 1940 and he continued the practice, begun when he was at the Admiralty, of firing off 'action this day' memoranda in all directions. That he must have been interested in the new tank seems to be indicated by the fact that it ultimately bore his name.

General Staff specification A22 became described officially as the Infantry Tank Mark IV, which indicates that it was fourth in the line of infantry tanks, all of which had been developed since 1935. The nomenclature of British tanks is horribly complicated and often a snare for the unwary. For example, although the term Infantry Tank Mark IV can be applied to any variant of the Churchill tank, each variant was also described as a mark in its own right. In the case of the Churchill, since they all featured the same engine these variants are identified chiefly by the armament and turret. Thus the first batch, armed with a 2-pounder gun in the turret and a 3in howitzer in the hull front were designated as Mark I, while those with a 2-pounder in the turret and a machine gun in the hull were Mark II. The Mark III indicated tanks with the 57mm (six-pounder) gun in the welded turret and a machine gun in the hull, while those similarly armed but with a cast turret were Mark IV. To add to the confusion, Marks III and IV also had 75mm guns fitted later in their

ANTI-TANK DITCH TEST

Crossing an anti-tank ditch – one of many tests carried out in the special test park adjoining the factory. Other 'obstacles' included shell-holes, sunken roads, steep gradients, water and deep, soft mud.

ABOVE A Churchill hull with fixed 'turret' being tested and demonstrated by Vauxhall. *(Author's collection)*

careers, and details of this 'upgunning' will be found later in the book.

Thus the full description of a Mark IV version would be A22, Infantry Tank Mark IV, Churchill IV. (Until 1948 mark numbers were always expressed as Roman numerals, only reverting to Arabic after that date.)

The new tanks were rushed into production, and it would be fair to say that the many failures suffered by the early models provide an excellent example of the old saying 'more haste, less speed'. In February 1941, when just two pilot models were complete and ready for testing, Winston Churchill issued an instruction that the new tank must be accorded priority 'in view of its superiority over all other models'; at the time this was a bold pronouncement given that the prototypes were proving to be mechanically erratic as well as causing confusion over the weapon it would carry.

The very early Churchills were unreliable and this is not surprising, given that they used a new engine and that the development of the tank had been so rushed. Such was the concern about their durability that, at one point, when invasion threatened, instructions went out that they should be parked up in suitable locations to serve as static pillboxes as a last resort. This, of course, never happened and the mechanical issues were recognised and addressed speedily.

One of the difficulties with the first Churchill tanks was the question of what guns to fit. It had been agreed, in committee at least, that

ABOVE **A very early Mark II (presumably, as there is no sign of the howitzer that would make this a Mark I, but it may be that the tank was unarmed), which was being put through its paces for testing. The photograph was released in September 1942 with the following caption: 'The Churchill Tank ready for action – The Churchill Tank, latest addition to Great Britain's armed forces, climbs a hill somewhere in England. In production for some time but on the secret list until now, the huge tank is so heavily armoured that it can be used as a pill-box.'** *(Author's collection)*

time. As an anti-tank gun the 6-pounder was an improvement over the 2-pounder even if tanks were not supplied with the high-explosive round initially. Still, it seemed, the role of the infantry tank was to fend off counter-attacking enemy tanks once a position had been gained. However, since the 6-pounder was as yet unavailable – and with the Prime Minister's imperatives ringing in their ears – the designers settled, for the moment at least, for the 2-pounder gun and a 3in howitzer in the front of the hull while stocks of these weapons lasted. Always remembering, of course, that by official decree the 3in howitzer was the close-support weapon in a tank, primarily there to fire smoke rounds.

The curious thing is that, although the 6-pounder was at least anticipated, an entirely new turret was designed for the 2-pounder gun in preference to adapting the larger 6-pounder turret to accept it temporarily. What is more, the turret provided for the 2-pounder was a casting whereas the turret then being designed for the 6-pounder by the Scottish firm Babcock and Wilcox was a welded structure with flat surfaces on every face.

Of course this is by no means the end of the Churchill story. An extensive rework programme and increasing crew familiarity resulted in a steadily improving performance,

the Churchill tank's turret should be designed to mount the new 57mm, or 6-pounder gun, although it was understood that this weapon was not yet ready, and might not be for some

RIGHT **The author in the commander's position in a Churchill Mark IV at the Tank Museum's annual Tankfest.** *(Author)*

and events in Tunisia and Italy showed that the tank had an astounding ability to clamber up steep slopes despite the relatively underpowered engine and archaic appearance. Many a German commander, who selected a defensive position on a high mountain ledge, felt his security ebb away as a Churchill edged up to confront him.

Subsequently, thicker armour and improved weapons made the Churchill even more formidable. All of which culminated, towards the end of 1943, in the appearance of the A22F, the Churchills Mark VII and Mark VIII. In essence these were entirely new tanks although similar in appearance and automotive components to the early Churchills and good enough to remain in service for some years after the war.

Many will remember the Churchill as the basis for various types of specialised armour, the so-called 'Funnies', which formed the spearhead of the attack on the Normandy beaches on D-Day, often manned by crews from the Royal Engineers. These tanks carried bridges, ploughed up mines and fired lethal demolition charges, and included the formidable Crocodile flamethrowers. Aside from its reputation as a fighting machine and the infantryman's friend, the Churchill tank also deserves to be remembered for these ingenious contrivances.

Improved versions of these specialist tanks remained in service long after the war, although as time went by, and new types of tank appeared, the service of the Churchill in the British Army gradually came to an end. Today they only survive in museums or in the hands of dedicated individuals prepared to spend untold hours, effort and funds to rebuild and run them.

ABOVE The author's Churchill heads into the Bovington Arena to meet the German Tiger for the first time since 1943. *(Author's collection)*

BELOW The beginning and end of an era. This photograph shows the first infantry tank, a Matilda, next to the last of the line, the Black Prince, which was proposed as the successor to the Churchill, but in fact never entered production. *(Tank Museum)*

Chapter One

The Churchill story

After its rush in to production, the Churchill was put through its paces in tests and in combat. As the lessons of evaluations and early battles were absorbed, a regular stream of improvements addressed its shortcomings and produced a tank that was to become the infantryman's best friend in battle.

OPPOSITE The author's restored Churchill Mark IV tank.
(Courtesy of Jonas Brane)

The origins

The Churchill's origins lie in the chaos prevailing at the start of the Second World War, and the lack of preparation that preceded it. Despite these inauspicious beginnings, the Churchill went on to defy the critics and become arguably the best British tank to have played a significant part in the war.

In early 1940 the Germans were some way ahead of the British in their study of tank tactics, design and manufacture. At that point Britain faced the challenge of creating and producing, from scratch, armoured fighting vehicles that would be effective in the new warfare that Germany's rapid advances had demonstrated to devastating effect. The First World War assumptions that had underpinned the Churchill's prototype, the A20, were clearly not going to be appropriate and time was short. A rapid and fairly radical rethink took place, leading to the specification of its successor, the A22, which it was hoped would fulfil a useful role in this new kind of fighting.

There had been little recent experience to go on when devising the tank, and at the best of times tank development is a race between the ability of an enemy to destroy it and the developers' ability to counter this threat while keeping the machine mobile and useable. The rush at the start of the war was far from the best of times to be doing this, and the Germans' capabilities had to be guessed at until they were met with in battle.

The minutes of the first Tank Board meeting on 24 June 1940, show that Prime Minister Winston Churchill made a very forceful point:

The Prime Minister enquired why alterations had been made in the specification agreed upon at the previous meeting [from the A20 to the A22]. The object was to get some kind of heavy tank in fair numbers early in 1941. No date could be given for production and a month had been spent in changing design. The Germans had built several thousand tanks since the beginning of the war, and we could not even produce 500.

The Prime Minister said that the choice was not between a good tank and a better one, but between a fairly good tank and no tank at all.

The rush to production and modification

Producing new tanks in wartime Britain was far from easy. Even with the design largely resolved, there was limited capacity and experience to undertake the manufacturing. As a consequence, the Churchill, as it was to be known, was made by a number of different companies in dispersed locations under the supervisory parentage of Vauxhall Motors. It wasn't just making tanks themselves that was

RIGHT A factory producing Churchill tanks. This line appears to be producing Mark II tanks, as the one nearest to the camera has a Besa mount in the hull, rather than a fitting for the 3in howitzer. *(Tank Museum)*

LEFT A small workshop making Petard mortars for the Churchill AVRE (see the assembled mortar on a bench in the middle of the picture, towards the right) as well as other equipment. *(Author's collection)*

BELOW A small workshop attached to the works making the Petard mortars, assembling components. *(Author's collection)*

difficult; a new engine had to be designed for the Churchill and this was done in the remarkably short space of three months, and that included making and testing a prototype. Through all of this, there were also shortages of materials and compromises that had to be made to enable production to get under way.

Unsurprisingly, many of the companies given the task of making these large machines were in the railway and shipbuilding industries. It was a labour-intensive process, and it would seem that in 1941 it took some 1,400 man hours to produce each tank.

Much of the work was further subcontracted to small firms that made components in workshops up and down the country. While this made development less vulnerable to bombing raids, it also reflected the fact that the main manufacturers were struggling to meet the demands of wartime production and adapting to a shortage of skilled workers.

As the tanks were rushed off the production lines, early examples were sent for testing by the Army, with engineers from Vauxhall on hand to turn their comments into modifications where possible. It is interesting to see how carefully these evaluations were carried out, even under the pressure of war. Many of the problems

RIGHT A woman grinding the base of a Mark VII turret, probably in the Babcock and Wilcox factory. *(Tank Museum)*

RIGHT A Mark I, T30972, being shipped out with its 3in howitzer removed. The mounting for the gun can be clearly seen on the right of the photograph (the howitzer fits through the larger opening). To the right is the housing for the sighting telescope. The ultraviolet lamp in front of the driver can be seen. This Churchill is about to be lifted by crane, hence the two cables slung under the hull. (Author's collection)

BELOW A well-known photograph of a Churchill Mark I. The 3in howitzer, with its muzzle cover on, can clearly be seen in the hull. The turret, with the 2-pounder gun, is traversed to the left in a position that would allow access to open the engine decks. The Bren gun is on a Lakeman anti-aircraft mount but has a standard magazine, rather than the 100-round drum magazine issued later with the tanks for anti-aircraft use. (Tank Museum)

that they identified were put right in a 'rework' programme as well as in ad hoc modifications.

One of these early Churchills, a Mark I with the registration T32246, was sent to Lulworth Camp in Dorset on 3 July 1941 to undergo trials. It was subjected to exhaustive testing and the resulting report gives a generally favourable view compared to other tanks that the testers were used to: 'Performance, so far as it could be tested, is good'; 'The steering is excellently light and effective and very little speed is lost on turning. Acceleration and manoeuvrability are impressive.'

However, there are detailed criticisms of certain items and comments on just how difficult it was to operate in some conditions: 'Dust was very prevalent during trials and this

RIGHT Mark II, *Jester*, T31693, with a cheerful Royal Tank Regiment crew. Note the effect of the mud spattered by the tracks, which have no track guards, and also the early downward-facing air intake louvres. (Author's collection)

RIGHT A Churchill Mark I, which appears to have been through the rework programme. It has the newer air intake louvres and full-length track guards, but still uses old, heavy, cast track. At the back is the drum-like jettison fuel tank. *(Author's collection)*

particular tank had no track guards or dust shields. As a result, it was essential to close down when on the move and most of the crew wore goggles – especially the driver and tank commander. During head-on shooting dust from the horns tended to obscure the gunner's vision. The muzzle blast from the 2-Pr. Gun blew dust in through the driver's roof flaps.' It was not long before track guards were specified to rectify this and other problems that came from running with uncovered tracks.

Other comments included: 'In general the stowage was good; well-arranged and left adequate space'; 'Comfort and roominess are very satisfactory.' However: 'There is insufficient height in the turret for a tall tank commander.' After the Marks I and II the turrets of all Churchills were noticeably taller, so it would seem that this point was heeded.

Internal, as well as external, illumination receives some interesting comments. In the era before infrared and later image-intensifying optics, tank crews had to contend with limited visibility

LEFT Inside the somewhat cramped turret of a Mark I or II. The loader on the right is ready to open the breech. On the left, just visible, the gunner looks through his telescope. Note how close the loader's/ operator's head is to the roof in this small turret. *(Tank Museum)*

LEFT The ultraviolet lamp to aid driving in convoy at night is on the left of the photograph, in front of the driver. The normal blackout lamp is on the other side. Inboard of both are the sidelights, which were later moved to the top of the track guards to give a better sense of the tank's width at night. *(Author's collection)*

NOT TO BE
STOWED
ON DECK

161

T31579

matter being available) and was clearly visible from the front at over 200 yards and could probably have been seen a mile away. It would appear to have little advantage over an ordinary lamp suitably screened.' A 1945 manual notes that the ultraviolet lamp will be replaced on later vehicles by a standard headlamp with a shutter-type front.

Visibility inside a tank when it is closed down is necessarily limited, and this greatly hampers the crew's ability to function. To deal with this in the Churchill, internal lights were fitted to the turret and driver's compartment, known as 'festoon' lights. The report points out that these internal lights were amply bright for the crew when the tank was closed down, but at night became far too easily seen from outside – being visible from as far as 150yd away, shining out through periscopes. The suggested solution was that the lamps should be fitted with rheostats so that they could be dimmed at night, and in due course this recommendation was adopted.

The periscopes were also tested to see how easily the crew could see out:

'When closed down and using the periscope over level ground the driver cannot see anything to his front nearer than 20 feet 6 inches from the front of his tank. This increases the difficulties of obstacle negotiation.' Tilting periscope mountings were produced to assist in this regard.

Another interesting report deals exclusively with trials that showed the damage caused to the outside of the tank by the muzzle blast from its own gun. One of the reasons that the front track-guard 'hoods' are missing in so many photographs of Churchill tanks in action is that the blast from the 6-pounder and 75mm guns tended to distort them. One measure that was taken to reduce the problem was the fitting of blast shields at the front of the track guards; however, these did not protect the hoods.

The rushed introduction of the tank meant that some 1,000 of the early tanks were produced with known and suspected faults. A key part of the Churchill story lies in the fact that these tanks were all subjected to a rework programme to incorporate improvements that Vauxhall and the Army recognised as being necessary from the early trials with the tanks.

at night. To address the difficulty of driving in the dark, particularly in convoy, each tank had an 'ultra-violet lamp', a headlight in front of the driver with a blue-painted bulb, which was meant to help illuminate the rear of the tank in front without the light being visible to the enemy. The manual optimistically suggests that: 'The rays from this light only reflect on any surface smeared with oil or grease or special paint, therefore the lamp is used for convoy driving, harbouring and entraining under blackout conditions.' The report notes that this was 'tested at night, and its performance was disappointing. It gave very faint illumination of greased sticks (no other fluorescent

The need for modifications to be made to so many tanks after production gave the Churchill the reputation of being unreliable. While this was true of the early tanks, it would be an unjust accusation after the rework programme was completed. As Vauxhall themselves said in the early handbooks: 'The defects exist solely because of the inadequate time that has been available for comprehensive testing. They are the "teething troubles" inseparable from a new design. In normal times they would have been eliminated – every one of them – before the vehicle was released for production. Times, however, are not normal. Fighting vehicles are urgently required, and instructions have been received to proceed with the vehicle as it is rather than hold up production.'

ABOVE Churchills under construction at an unknown factory. The numbers chalked on the front of the hull nearest the camera suggest that it would be T31196. Behind is T31035. The figures R625 on the first, and R626 on the second tank suggest that the work being carried out at this stage (fitting track guards) is part of the rework programme. This could account for the fact that while the tank's 'T' numbers are not sequential, the 'R' numbers are. *(Tank Museum)*

The rework was successful in reducing, if not eliminating, many of the early problems and in producing an outstanding tank. This was amply proved during the fighting in North Africa, which largely silenced the early doubters and laid the basis for the tanks' fighting reputation.

Once the rework programme had addressed the main issues that could be foreseen from testing and trials, the Churchill's future evolution

ABOVE This Mark I tank is preserved at Bovington and shows the early air intake louvres and the early, heavy, cast track. *(Author)*

Following early testing of the Churchill, a full list of modifications that were regarded as essential was compiled. Key changes which affected mobility and fighting characteristics of the tanks included:

- Waterproofing of hull seams with 'Bostik' waterproofing cement to prevent water seeping in when wading or crossing streams.
- New air louvres with the inlets facing upwards in place of the initial design that faced down and drew leaf litter and dust into the engine.
- Full-length track guards to stop dust and debris from coming on to the decks and impairing vision.
- On newly manufactured tanks, better air flow was provided through the engine compartment by enlarging the ventilation hole in the back of the hull and extending the space through which it could exit from 5in to 8in. This change can be seen by looking at the number of slats in the space behind the gearbox compartment.
- Fitting blast shields to the front and rear sections of the track guards to prevent them from being distorted by the muzzle blast from the main gun.
- Installing a ventilation door in the bulkhead, which separated the back of the fighting compartment from the engine, to allow gun fumes to be drawn back into the engine compartment.
- Fitting rear smoke dischargers.
- Changing gun mountings to reduce the tendency for incoming machine-gunfire to cause 'bullet splash' past the Besa and main gun mountings.

Many other modifications were listed, improving just about every aspect of the tank including suspension and tracks, final drives, front idlers, brakes, engine, clutch, radiators, petrol tanks, gearbox, controls, electrical equipment and instruments, traverse mechanism and stowage. It is a very comprehensive list and the changes range from newer materials where previous ones had been found to fail, to strengthening components and, in some cases, redesigning items that had caused problems.

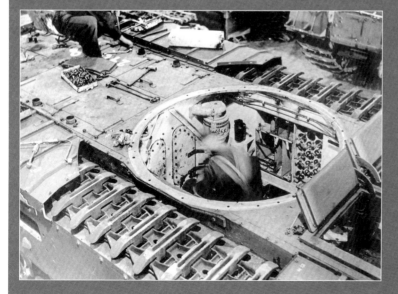

LEFT This Churchill is almost certainly a subject of the rework programme. The man appears to be working on the bulkhead ventilator door, which is at the back right of the fighting compartment. The fragile clips and ammunition stowage are visible on the right in the fighting compartment area. Each clip holds in place the round above and below it. *(Tank Museum)*

came in response to the reports of those who fought in the tanks and the issues that they encountered. In its short production life the Churchill was adapted and modified to reflect these lessons as they were learned on the beaches at Dieppe, through North Africa and Italy. They helped to transform the tank into a better fighting machine, not least in the form of the heavily armoured Mark VII and the many variants prepared in time for the invasion of Normandy. Just as importantly, they improved the tactics used by Churchill regiments, as painful experiences taught the crews how to make the best use of their tanks.

Battlefield lessons: a race to keep up with the enemy

The real test of any tank is how it performs in action and it was not long before reports were coming in from the battlefield rather than the testing area. The fighting in North Africa provided a vital proving ground for the tanks, as it showed at first hand what were to be their strengths and weaknesses. For example, a 1943 report notes that 'it is clear that the 2 Pounder gun is no longer useful, and that there is a need for good armour piercing and high explosive ("H.E.") ammunition' (the 2-pounder was fitted to Marks I and II).

As the 3in howitzer was fitted to the Mark I Churchill, and 6-pounder guns were fitted to the Mark III tanks that fought alongside them in North Africa, the report is able to comment on their dual effect: 'The combination of 3 inch Howitzer with 6pdr and Besa machine gun was effective, especially when, at longer ranges, H.E. was used to destroy tank crew who had baled out and gone to ground in hollows.'

One particular concern was that the Churchill's turret had an internal mantlet rather than having the mantlet on the outside, as on the Sherman. The report says: 'The mantlet was very vulnerable to hits from 75mm and 88mm guns and hits on this component frequently caused the mounting to break away at the trunnions. An external mantlet might help to deflect shots and strengthening of the trunnions should lessen the danger of mountings breaking away.' This

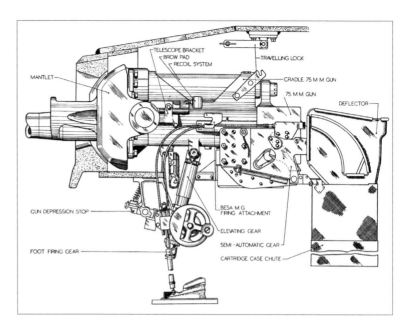

problem with the mantlets led to a remarkable modification of some Mark IV Churchills to the NA 75 version, by taking the external mantlet of damaged Sherman tanks and fitting it to the Mark IV Churchill turret. (See page 60 for more on this conversion.) However, the 'official' Churchill retained its internal mantlet to the end.

The ability to provide smoke cover both for the tanks themselves and for the infantry was a concern, and in the absence of smoke rounds for the 2-pounder and 6-pounder guns, the crews relied on a short-range 2in mortar fixed to the turret roof, referred to as a 'bomb-thrower'. The report notes: '2" Bomb-thrower – instantaneous smoke was necessary, also greater range.' Both these points were quickly corrected, the first by issuing bursting white phosphorous ammunition and the second by using a stronger, 55-grain, propellant charge in conjunction with a facility to adjust the range of the bomb-thrower.

As for gunnery, it is interesting that, despite the limited effective range of the guns, the 1943 report notes that in the sighting telescope 'cross wires were too thick, frequently obscuring the target at ranges of 1,000 yards and over. Slightly increased magnification and illuminated cross-wires for night operations were also required.' Both were subsequently provided as a result of these comments from the battlefronts.

A recurring criticism – and cause of casualties – was the lack of armoured stowage

ABOVE Mantlet and gunmounting showing the position of the internal mantlet inside the turret. *(Author's collection)*

ABOVE This photograph illustrates the differences between the smoke rounds for the 2in bomb-thrower. At the top is a bursting white phosphorous round; in the middle, with its fins slightly unscrewed, is a base-ejecting early round; the bottom round is also a bursting kind, and behind it is the 50-grain propellant charge that is inserted inside the fin/base section. Note that the top round has pressed steel fins, while those of the other rounds are made from curved metal alloy. *(Author's collection)*

BELOW A 2in bomb-thrower of the early type fitted to the roof of the Mark IV tank. *(Author's collection)*

RIGHT This is the later type of bomb-thrower, which loads from the base. It is fitted inside the Churchill Mark VII at Bovington. To the left is the armoured stowage for the mortar rounds. The front door is off the stowage box, revealing the centre tray for ten rounds of phosphorous ammunition. The tray can be pulled out by the little 'D'-shaped handle at the base and jettisoned if there is a fire inside the tank. *(Tank Museum)*

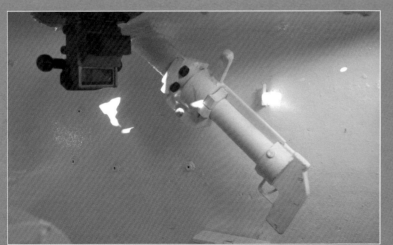

MAKING SMOKE

The Churchill had various methods of providing smoke cover for itself and for the infantry it fought with.

First, the larger guns in the tanks – the 3in howitzer, the 95mm close-support howitzer and the 75mm – all had smoke ammunition to provide a screen at a distance.

Second, there was the 2in mortar in the turret roof referred to as a 'bomb-thrower'.

The bomb-thrower came in a variety of marks and was fixed to the turret roof at an angle. Early marks had a pistol grip and trigger. This version of the weapon opened much like a shotgun, breaking on a hinge just below the turret roof so that the loader/operator could drop a mortar round into the back section and lock it closed, ready for firing. Later variants of this bomb-thrower loaded from the base and enabled the range of the weapon to be varied by venting gas out through a gas regulator to reduce pressure and thus range.

Two types of ammunition were carried: one emitted smoke from the base of the round. The other used a bursting phosphorus composition for more immediate effect.

Mortar rounds that contained phosphorous

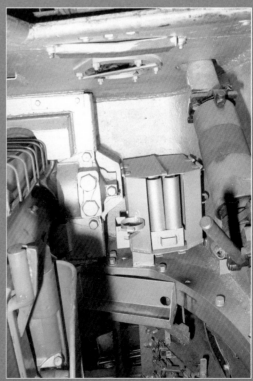

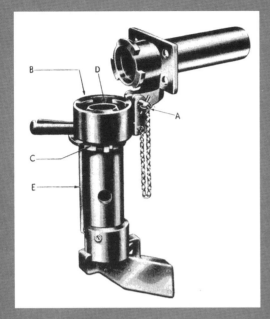

LEFT The early bomb-thrower opened for loading. (*Author's collection*)

ABOVE A 2in bomb-thrower inside a Mark VII. This image shows the back of the weapon open and ready to be loaded. Note that this tank was originally painted white inside but was repainted in silver at some point after the war (the over-painting can be seen on the WS19 Set junction box on the right of the picture). (*Tank Museum*)

ABOVE This picture shows the smoke grenade holder on the rear of a Churchill Mark IV. Next to it, with its lid open, is the junction box to which the wires from the two grenades would be connected. Once wired in, the driver could set off the grenades by means of switches beside him – one for the left hand side and one for the right. (*Author*)

were extremely dangerous; the body of the round was made from thin steel and they could be ignited following a hit on the tank, even one that did not penetrate. Such hits frequently sent flakes of metal flying around the inside the tank. In the pre-Mark VII tanks the mortar rounds were stored at the front of the turret in a flimsy tin with no cover. The Mark VII tanks (and Mark VI tanks with armoured stowage) had tougher, all-enveloping stowage for the mortar rounds that protected against flying flakes and delayed the speed with which a fire inside the tank would affect the rounds. Burning phosphorous caused very serious, potentially fatal, wounds.

Third, although not weapons as such, the tanks also carried 4in smoke grenades or 'emitters', which were fitted in pairs in carriers on the right and left rear of the tank. These could be set off electrically by the driver and allowed the tank to leave a trail of smoke behind as it moved. A later version enabled the grenades to be dropped from the tank after being lit.

RIGHT The restored Churchill Mark IV standing next to an appropriately named Challenger Main Battle Tank. The two smoke grenade holders can clearly be seen. At the time of this photograph the junction boxes (as well as the tow-hook) had yet to be fitted. (*Author's collection*)

RIGHT The loader/operator stands in this small area. The upright four-round stowage bin that is used as a step can clearly be seen. Behind the main gun is the spent cartridge catcher and below it the canvas chute into which the spent cases fall. *(Author)*

ABOVE This is the protective clip on the base of a 6-pounder round. The clip can quite easily be knocked off. *(Author's collection)*

RIGHT Here is the ammunition storage 'wine rack' on the loader's side, which holds 39 rounds. On the left side of the rack are clips for holding additional rounds of ammunition by the pannier door, which can be seen in the drawing with the number '15' in the opening. These are the rounds that often caught fire. *(Author's collection)*

for ammunition. This is an extract from a report by Peter Gudgin, of whom more is said later:

1) Armouring of 6pdr ammunition bins is essential. … Ammunition fires in my own tank were caused apparently by hot splinters penetrating the bins.
2) Grenade rack should be moved from its present vulnerable position beside the commander to one more protected from hits on the turret.
3) Alternative armoured stowage for the four 6pdr rounds behind the driver is necessary. These were a frequent starting-point for ammunition fires.
4) The upright ammunition bin near the operator must have a lid.

This last comment may need a little explanation. In Marks III, IV and VI (before its conversion to armoured stowage) there is an ammunition bin at the back of the fighting compartment just under the loader's hatch, which holds four rounds of 6-pounder ammunition, and after the conversion of the tanks to 75mm guns, three rounds of that ammunition. As a report on Operation Trent put it: this 'may be the cause of serious accident unless a cover is provided for the top of the rounds (which are stowed projectile downwards). This rack is a very convenient foothold for getting into the turret. The clips protecting the percussion cap are frequently knocked off, and personnel entering the tank are liable to step on the rounds and possibly set it [sic] off.' One reason this danger existed was that the crew were likely to be wearing hob-nailed boots!

The standard ammunition stowage for the main gun – in what resembled 'wine racks' in the sides of the fighting compartment – was also tricky. Each round had to be prevented from falling out of its tube and for this a fragile spring-loaded clip was used.

In the report on Operation Trent there are some critical comments on this clip: 'When ammunition is not fitted, it [the clip] is liable to be broken off. When the ammunition is in place, the spring which controls the position of the

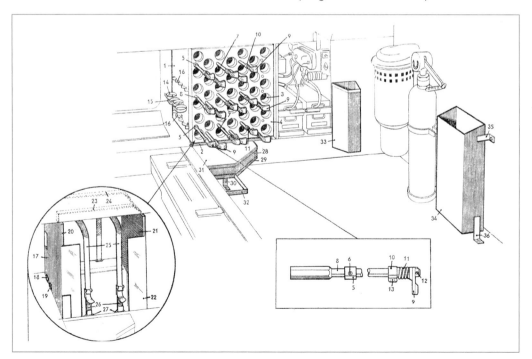

clips securing the rounds does not always hold the clip in the correct position, and the rounds work out and jam the turret.'

The ammunition was stored in cardboard tubes referred to as 'cartons'; the same report notes that 'rain gets in and soaks the cartons in which the ammunition fits and swells them up. The rounds cannot then be withdrawn.'

The danger of fires in tanks was always present in the minds of their crews and of the designers. The fires that the crews most feared were caused by ammunition, not petrol, and the primary source of the fire was the propellant in the brass shell cases and the phosphorous in the mortar rounds – all of which were stored in the fighting compartment alongside the crew. As a direct consequence of these experiences, all of the Churchill Mark VII and the modified Mark VI tanks had armoured ammunition stowage for the 75mm ammunition and enclosed boxes for grenades and mortar rounds. They also did away with the upright stowage bin behind the loader/operator.

Ventilation in the Churchill under battle conditions was always bad, but worse in the early versions where the battle report notes that it 'places excessive strain on the crew. When the engine is not running, as frequently occurred in long spells of "line-holding" from hull-down positions, and the tank is closed-down, due to

LEFT Early, open topped, stowage for 2in mortar rounds. This particular example is in the turret of a Mark IV. The turret front is just beyond the tin. To the left is the trunnion holding the gun. *(Author's collection)*

LEFT Armoured stowage for 2in mortar rounds in the front of a Mark VII (this tank is in the Cobbaton Combat Collection in North Devon). The centre section of the box, behind the plate with a clip at the top, lifts out. The idea was to keep the phosphorous rounds in this part so that it could be jettisoned in the event of fire. Clipped to the front is a Wesco oil tin with flexible spout and the canvas item on top of the box is the kit of tools for the Besa machine gun. *(Author's collection)*

RIGHT On the left is the turret extractor fan inside a Churchill Mark IV. It is the old 'snail' type with a hose (shown disconnected) to enable it to draw fumes from over the Besa coaxial machine gun. In later castings of this turret, the whole fan assembly moved to the right. On the roof to the right of the fan hose is the canvas case for stowing maps. Below that is the recuperator for the main gun. *(Author's collection)*

RIGHT The later type of extractor fan (seen fitted to a Mark VII) was placed in a better position to draw fumes from the Besa. A festoon light is just to the right of the fan, and the chain to the left can be unhooked to hold up the top cover of the Besa when reloading. The recuperator for the main gun is at the base of the photograph. *(Author's collection)*

accurate mortar-fire, it is practically impossible to breathe after an hour without opening a hatch, less if the firing has been heavy. Both the hull gunner's and the turret fans are inadequate, both for carrying away heat and fumes.'

Improved fans were fitted as a consequence of this observation, and their position in the turrets of Mark IV, V and VI tanks was moved to the centre of the roof to be more effective. Additionally, a small door was installed in the bulkhead at the back of the fighting compartment, which could be opened to allow the engine cooling fan to draw air out of the turret. Even these changes did not really provide full relief.

Another perennial problem came with the track guards. The early reports had stressed how necessary these were; however, fitting them brought new problems: 'The design of the track guards is unsatisfactory. Being of light sheet metal, the guards are readily damaged by the lightest attack and when this occurs on the centre section, the fringes of ragged metal which are thrown up by oblique penetrations prevent the turret from being rotated.' This difficulty, and the trouble caused by mud packing under the track guards, meant that many crews removed the centre sections to

avoid the turret being jammed and unable to traverse; or worse, lifted up so that the traverse gear broke.

The vision arrangements were found to be inadequate, and the suggestions for improvement show how different fighting conditions taught different lessons. The early stages of the war had mainly seen tank warfare in open desert conditions. In the closer countryside of Tunisia, and even more so in Italy and north-west Europe, danger could lurk at close range and in almost any direction. The report notes: 'Points are coming to light which did not apply in the Middle East where tanks were not closed down and commanders frequently exposed their heads to get good vision. Designers have always understood that a gunner's periscope [as opposed to his sighting telescope] was in the nature of a luxury to enable the gunner to get a general view of what was happening when he was not actively engaged in operating his guns. This is not however the case. He has a definite part to play in spotting for targets over a definite zone.' This comment reflects the fact that, oddly, the standard turret on the Marks IV, V and VI did not have a gunner's periscope, although all other marks before and after did.

BELOW A Churchill Mark III or Mark IV AVRE, *Scimitar*, with the centre section of its track guard missing to avoid jamming the turret. Note also the plentiful additional armour made from spare track links. This tank has an extended driver's periscope and is using old heavy cast track. The tank behind is carrying a large fascine bundle with the commander on top to direct the driver.
(Tank Museum)

'Similarly the loader covers another zone while the commander covers the remaining countryside. The commander has also to watch neighbouring tanks. To do this it is usual for the commander to swing the cupola continuously to and fro, while looking through his periscope as the tank advances. This is a very tiring operation unless the cupola race is very free and this is not always the case in the Churchill … it is generally felt that a raised cupola giving all round vision is the correct equipment for the commander.'

These tanks were using the early, two-periscope, commander's cupola – this problem was recognised and cured with the later 'all-round-vision cupola'. In Tunisia, the North Irish Horse, with typical inventiveness, dealt with the problem by fitting captured German all-round-vision cupolas to some of their tanks.

'The driver's vision arrangements are not satisfactory. When using the periscope he cannot see to either side and since the other members of the crew are engaged in their normal duties when closed down they cannot warn the driver of the presence of vehicles on either flank. Some collisions occurred in the UK due to this. It is felt that the driver's periscope should be raised to give a clear view over the horns of the tank. Users agreed that the driver's visor was inclined to be a weakness in design and that it was difficult to use. They would agree to

its abandonment and to the provision of two periscopes for the driver.'

All of these comments were attended to and incorporated in the Mark VII/VIII Churchill in which the driver got two periscopes and the gunner, one.

The Churchill vindicated

The faults in early Churchills *before* the rework programme had made some in Britain doubt the wisdom of continuing with production of the tank. However, those involved in the fighting noted that the reworked Churchill was as effective at raising the spirits of the men who fought in and with it as it was in

dampening those of the enemy. Here is a well-earned tribute from those who knew how the tank had performed.

'The Re-Work is in an entirely different class of reliability, as compared to its predecessors. Gearbox, clutch and suspension which gave so much trouble previously now give practically none.'

Another report affirms:

'So far in the Tunisian campaign the Churchill has done very well. … The tank itself has been used against many different types of opposition and has acquitted itself well. The morale of the crews is high, the presence of the tanks gives confidence to the infantry and all users are agreed that it would be a great mistake to stop production of the Churchill until a tank with equal or better protections and with a better performance has been produced and proved successfully.

LEFT A Churchill Mark IV or VI of the King's Own Regiment, which has driven over a mine. There is an ARV in attendance (it is behind the gun tank). The fitters are helping the crew. The two on the right are adjusting the track by moving the front idler with special spanners carried for the purpose. The track is broken at the back. A fitter is standing on the track: note the missing centre section to prevent the turret jamming. The front hoods are missing from the track guards and the blast shield on the left of the picture is bent inwards. There are two links of old style heavy cast track above the driver's visor to be dropped over it as additional armour. The tank has an early cupola and blade vane sight as well as a raised periscope for the front gunner. *(Tank Museum)*

'The reliability of the reworked Churchill is best illustrated by the fact that, due to operational necessity, 22 tanks carried out a march of 70 miles without any serious mechanical defects while 9 of the vehicles went on to complete 103 miles in 24 hours.

'The tanks are being used in a somewhat unorthodox way in order to assist in holding an extended forward position. In a few cases they have held forward defence lines which were too exposed for the infantry to hold. During this time they have been subjected to mortar fire for considerable periods and have stood up to it well.

'In order to give increased protection to the sides of the tank users had begun to attach spare track links to each side between the air louvres and the escape doors.'

This practice continued through the war, and it seems that the official view was that it did not really increase the protection of the tank but greatly helped the morale of the crews. The real solution came with the addition of increased armour thickness on the Mark VII tanks.

The report also noted that Besa machine guns were 'extremely popular and all users consider it to be outstandingly successful'.

Another account from Tunisia records that Besa fire on a German anti-tank gun penetrated the shields of the gun and killed the crew. These guns were retained on the tanks throughout the war.

The report also says that: 'The use of Churchill as an armoured recovery vehicle has been very successful and users are suggesting that the establishment of these vehicles should be increased from two to four per regiment although some units would like still more.' The armoured recovery vehicle or ARV variant of the Churchill went on to become an indispensable part of its role in the Second World War and also in Korea.

Thus the experiences of the early phases of the war were documented and then put into practice; the result was a series of adhoc modifications to the Churchill, many carried out in the field, but more fundamentally they led to the evolution of the Mark VI , Mark VII and VIII versions of the tank and, just as importantly, to a clearer understanding of the best ways in which to use these tanks (see page 145).

The most significant change came with the Mark VII tank – referred to sometimes as the 'Heavy Churchill' – which was, in many ways, a new tank. It had similar frontal armour thickness to that of the Tiger tank, albeit with an inferior gun. Despite these changes, this machine was not significantly heavier than its predecessors although it was a little slower. Equally importantly, other lessons had been put into good use with the development of variants of the tank that were to have specific roles in the Normandy landings and the fighting in Italy and north-west Europe. Having seen how difficult it was in Dieppe and elsewhere to tackle strong fortifications, the developers created the AVRE with its very potent, but short range, petard mortar and Crocodile flame-thrower version of the Mark VII bridge layers and ARKs, as well as variants to lay matting across shingle. Many of these are described in Chapter 2.

There is a useful summary of the Churchills' performance after the start of fighting in Normandy in the War Cabinet's response to a critical report by the Select committee on National Expenditure. The Committee had sought to blame various departments for perceived failures in tank production. The reply came from the Prime Minister himself, and the extracts below relate to the Churchill that bore his name. The reply is dated 2 August 1944 – not very long after the fighting in Normandy had started – and contains a warning: 'There are certain aspects of future development, such as the new special 6-pdr shot [the discarding sabot armour-piercing round] which must be treated with the greatest secrecy, and I feel sure that your Committee will realize that fact.'

The Prime Minister cited a quote from General Leese, commenting on the fighting in Italy:

'It may interest you to know of the fine performance of the Churchill tanks which supported the Canadian Corps when they attacked and broke through the Adolph [sic] Hitler Line last month. The Churchills stood up to a lot of punishment from heavy anti-tank guns. Several tanks were hit hard without the crews being injured. They got across some amazingly rough ground. Their 6-pdr guns made good penetration and were quick to load and aim.

'So good was the work of the Brigade with the Churchills that the Canadians have as a privilege asked them to bear the Maple Leaf on their tanks.

'The tank crews have come through their successful attack on the Adolf Hitler Line with tremendous confidence in the Churchill tank. We shall make good use of every Churchill you send us.'

The reply goes on:

'It is perhaps too early to come to a final judgment on matters of relative efficiency

of different types of tank in the light of conditions which are being and are likely to be, met with in North West Europe. But, whereas in previous operations it cannot be denied that, with the exception of the Churchill tank, the most effective tanks and tank weapons available to the Allies have been of American origin, the lead both in armament, ammunition and armour protection has passed to tanks of British manufacture and design.'

Near to the end of the war, between 25 September 1944 and 26 October 1944, when the evolution of the Churchill as a gun tank was effectively complete, the Australian Army tested three Churchills, a Mark IV, Mark V and a Mark VII, as well as two Shermans, a M4A1 and a M4A2 in New Guinea. The test provided a direct comparison between the tanks and a (presumably dispassionate) view of which was preferred. The extract below comes from the 'Preliminary Report on Operational Trials of Churchill and Sherman Tanks' in the Tank Museum archive:

'During the period in which the vehicles were in New Guinea they each covered an average of 130 miles all of which were run in first and second gears owing to the heavy 'going'.

'The following advantages of the Churchill tank would favour their use in preference to Sherman tanks:
i   Superior manoeuvrability, particularly at slow speeds
ii   A more suitable low gear ratio for infantry co-operation
iii Greater thickness of armour
iv   Slightly better performance in creek crossing and climbing
v   Greater ground clearance.'

The different marks of Churchill

C hurchill tanks went through a series of major evolutionary changes reflected in the different Marks I to VIII (not to be confused with the overall designation of the tank as an Infantry Tank Mark IV).

TANK REGISTRATION NUMBERS

Small changes were made throughout the life of the tank, and these can be seen by a suffix after their War Department number – the number, beginning with 'T', which was, in effect, each tank's registration number. This remained with it for the duration of its service, even though it might be allocated to different units and given other names by them. (Tanks that remained in service after the war were then renumbered with a new system, comprising numbers and letters.)

The Churchills' 'T' numbers would be followed by suffix letters (any of R, A, B, C, D or E) depending on what was done and by whom. For example, all vehicles that had been through the rework programme should have had the letter 'R' added. Later in the war, once appliqué armour was fitted to the hull sides of Marks III and IV, all tanks so

modified were meant to have 'E' after the number to signify that all modifications had been carried out.

LEFT Churchill Mark IV, T31699R. The 'R' suffix shows that the tank has been through the rework programme. The wooden block attached to the blast shield on the right of the photograph is a 'jacking block', used to place under the hydraulic jack on soft ground. *(Tank Museum)*

The main differences between the marks are as follows:

Mark I – Standard square door hull with small cast turret mounting a 2-pounder gun (Ordnance, Q.F. 2Pr., Marks IX, IXA, X or XA) in the turret with coaxial Besa (Besa 7.92mm. M.G., Marks II, III or III*) on the right of the 2-pounder gun; 3in howitzer (Ordnance, Q.F. 3in Howitzer, Marks I or IA) in the hull.

A few examples of these early tanks existed with 3in howitzers in the turret and 2-pounder guns in the hull, and at least one with 3in howitzers in both. In fact, placing the howitzer in the turret made more sense; it was a low-

LEFT The 3in howitzer in this Mark I has been moved to the turret, and the 2-pounder gun is fitted in the hull. This tank has early track and air intake louvres. *(Tank Museum)*

velocity weapon that needed more elevation than the front hull mounting could allow, in order to be effective at longer ranges. The 2-pounder, by contrast, was a relatively high-velocity weapon that would do most of its work as an anti-tank weapon at ranges where a reduced amount of elevation was needed.

Mark II – Same turret as Mark I but with a Besa in place of the hull-mounted howitzer.

Mark III – Same hull as Marks I and II. New welded turret with round pistol ports in the rear sides, mounting a 6-pounder gun (Ordnance, Q.F. 6-pounder, 7 Cwt, Marks III or V) in most examples, although also mounting the 75mm

ABOVE This Churchill Mark II, *Penelope*, is equipped with early air louvres facing downwards. It is noteworthy that this tank has been fitted with full track guards, although its number does not have the 'R' suffix to suggest it has been reworked. The tanks behind do not appear to have track guards. *(Tank Museum)*

RIGHT A Mark III with appliqué armour (note the two squares on either side of the gun) but running on old, heavy cast track. This tank is armed with a 75mm gun. *(Tank Museum)*

gun later (Ordnance, Q.F. 75mm, Marks V or VA). The two variants of the 6-pounder were a short-barrelled Mark III version of the gun in earlier tanks and a higher-velocity, long-barrelled Mark V gun in later ones. These guns were initially 'free elevation'; that is to say the gunner used a shoulder piece to move the gun up and down. Some later guns – probably most or all of the 75mm ones – had geared elevation. The coaxial Besa (Marks II, II*, III or III*) moved to the left of the main gun, causing some inconvenience as the ammunition box was stowed on the other, right, side of the main gun and the belt fed through below it to the left

ABOVE **An early Mark III, newly reworked. From the next photograph we know that this is T31272R, in pristine condition, with a short-barrel 6-pounder gun. The tank has an additional fuel tank on the rear. Just ahead of the fuel tank is the space for warm air to exit from the engine compartment (note that this air-venting arrangement has only two slats, in contrast to the larger opening on the Mark VI hull). The turret has the first type of sighting arrangement for the commander, a simple metal stake, which preceded even the blade vanes.** *(Tank Museum)*

LEFT **T31272R, seen from the front with new track guards, new air intake louvres and (seemingly) new tracks.** *(Author's collection)*

ABOVE **Mark III, T32149R, with long-barrel 6-pounder.** This tank has been fitted with a fascine carrier (just visible under the gun barrel) and has dropped the fascine bundle which it has used to help climb the wall. The pistol port at the back of the turret is open because the cable securing the fascine bundle would have been released from inside the turret. Note that, despite being a reworked tank, there are no track guards. The cover on the air louvre is an unusual one; these were often made in field workshops to local design. *(Tank Museum)*

ABOVE **Churchill Mark IV, T68421R,** used for training. The loader/operator has the microphone for the WS19 in his hand. Standing behind him outside the turret is the instructor, whose right hand is holding on to the aerial base for the WS19 'B' Set. The tank has a very basic blade vane sight in front of the commander. Interestingly, the tank appears to have been painted with a camouflage pattern. *(Tank Museum)*

where the gunner sat. From this mark onwards, the Ordnance types remained the same until the earlier 6-pounder gun was withdrawn from service.

Mark IV – Same hull as Marks I through III, but with a new, cast, turret with square pistol ports in the rear sides, mounting a 6-pounder or 75mm gun.

Mark V – Same hull as Marks I through IV, but with a 95mm tank howitzer Mark I in the turret with geared elevation.

Mark VI – Essentially the same as a Mark IV, with probably only 75mm geared elevation guns, which were produced from late 1943 onwards (acceptance trials being in August of that year). The Mark VI had a larger outlet for air at the

RIGHT **A pristine Mark IV (or VI),** showing a rear-mounted additional aerial for the WS38 and a raised driver's periscope just visible over the track guards. It appears to be fitted with a spare track link of the old Dieppe-era heavy cast type. This tank has a 75mm gun. *(Author's collection)*

LEFT A clear photograph of a reworked Mark IV with a short-barrel 6-pounder gun. Note the small fire extinguishers on the back decks and the spare (early, heavy) track links on the engine hatches. There is a jettison fuel tank at the rear of the hull. The turret has the mushroom air extractor cover on the front left; it was later moved to the centre to let the fan sit above the Besa. There is also a very early sighting stake on the front of the turret. Note also the small air exit on the back, and what look like new air louvres. Given the presence of new, upwards-facing louvres, the complete track guards and the 'R' suffix on the tank's number, it is likely that this Churchill has just been through the rework programme. *(Tank Museum)*

BELOW A Churchill Mark VI, T252227, named *Saturn*. Note the greater number of slats in the rear of the tank to improve air flow, the air vent in the centre of the turret roof and the 'U'-shaped strip of armour bolted as a protective plate on the turret base (above the 'urn' of *Saturn*), on the right hand side. The turret also has a mounting for a PLM anti-aircraft mount (the circular fixture between the fronts of the hatches). This was designed to allow use of a pair of Vickers K guns, mounted on the turret roof and fired from inside the tank. However, no evidence can be found of the guns being used like this in action. *Saturn* has no suffix to its 'T' number which, if present, would have denoted rework. It has a 75mm gun with an additional counterweight just behind the muzzle brake (which could indicate free elevation). The towing cable is stowed in a rather unorthodox way, but this can also be seen in other photographs of Churchills. Aside from the long crowbar clipped to the track guard nearest to the camera, the other stowage items such as shovels, pick axe and sledge hammer, which should be on the back, are missing. *(Tank Museum)*

RIGHT A clean Mark V with 95mm howitzer, raised driver's periscope mount and additional aerial at the back of the turret for the WS38 Set. Note that the spare track link is an early cast type. *(Tank Museum)*

back, but it is difficult to be certain that this was not also fitted to late versions of the Mark IV. Also, the Mark VI was later provided with armoured ammunition stowage similar to that in the Mark VII, save that the bin on the turret floor was larger on the Mark VI and had a lid.

Mark VII, the A22F – A different hull and turret, the latter cast. This tank had significantly thicker armour and a reworked gearbox giving a slightly reduced top speed. The Mark VII appears to have been sent for acceptance trials from late 1943.

Mark VIII – Same as Mark VII but mounting the 95mm close-support howitzer.

DIFFERENCES IN TURRETS ON THE MARKS IV, V AND VI

BELOW Mark IV, T68831R, has been reworked and has the later turret with central extractor fan. Just by the right hand of the man in the turret is the fitting point for the PLM anti-aircraft mount. This tank is being tested in a wading tank at the designated 'fording depth', which was often marked with a line on each side of the hull. The driver's periscope holder is not fitted and neither is the main gun, which allows the internal mantlet to be seen clearly. The lorry in the background is carrying two Churchill engines. *(Author's collection)*

As will be seen from the photographs on pages 34–7, there are some differences between the cast turrets found on these marks. The main changes are only visible on the top, and involve the position of the 'mushroom' that covers the ventilation fan. On early turrets, this was near the left side, in front of the gunner. These tanks used a 'snail' fan with a pipe extending over the Besa machine gun to extract fumes. The turret fitting for this fan is incompatible with the later, more powerful type of extractor fan, which was fitted more centrally on the turret. One difference between the turrets is visible from ground level, on the right-hand side. Later turrets not only had the different fan position on the roof, but also had an armoured plate bolted to the outside of the turret on the right to protect the turret ring, suggesting that the casting must have been altered (see photograph on page 35). Late versions of the turret also have a fitting for the WS38 aerial cast into the rear of the roof. Some have a fitting between the commander's cupola and the loader/operator's hatches for the PLM mount, for twin Vickers K guns to fire at aircraft.

LEFT An early Mark VII, probably T173143 H. Note the absence of bulges on the turret front, the early cupola and blade vane sight. The tank has a small triangle bolted to the turret front, which may indicate that it was an unarmoured prototype. *(Tank Museum)*

LEFT A Churchill Mark VII. This tank has a later version of the infantry phone on the back, an all-round vision cupola and provision for the WS38 aerial at the back of the turret. *(Author's collection)*

LEFT A Mark VIII in good condition. It is not clear how many of this version saw action. *(Author's collection)*

SHOOTING AT AIRCRAFT

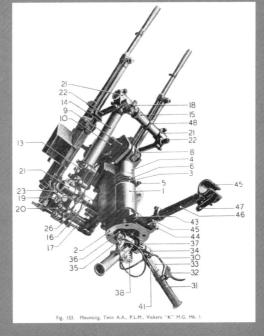

Fig. 153. Mounting, Twin A.A., P.L.M., Vickers "K" M.G. Mk. I.

ABOVE Mark II, T32092, with the Lakeman mount for the Bren gun. The tank also has a spotlight behind the Bren. *(Tank Museum)*

RIGHT This is the PLM mount for a pair of Vickers K guns. At the base are two handlebars with controls for aiming and firing the guns. These would have been located inside the turret while the plate (2) would have been fixed outside to the turret roof. *(Author's collection)*

RIGHT A Churchill Mark IV under restoration in Belgium. Note the small mushroom-shaped cover between the loader's hatch on the left and the commander's cupola. This covers the PLM mounting point. *(Author's collection)*

Early in the war it was evidently decided that tanks needed a means of protecting themselves against air attack. They carried a Bren gun with an unusual 100-round drum magazine for this purpose and initially also had a frame to mount it on, called the Lakeman mount after its inventor, Major Lakeman. The mount was stowed on top of the turret when not in use and was basically an arm from which to hang the Bren gun (see photograph on the left). However, to use it and fire at aircraft the commander had to half emerge from the turret, fit the mount and Bren gun and then open fire. As is noted at page 49, more considered advice on dealing with aircraft was to remain closed down – to avoid the nearly suicidal risks of being halfway out of the turret while under fire. In any event, the Lakeman mount was a temporary expedient before an even more extraordinary anti-aircraft solution was provided – the PLM mount. This fitted to a plate installed in the centre of the turret roof. The mount held a pair of Vickers K .303in guns (once used in aircraft, and by this stage no longer required by the RAF, and famously also used on Jeeps by the SAS when raiding). These were controlled from inside the turret by a pair of handlebars. The whole apparatus must have been very clumsy and, aside from the difficulty of mounting and dismounting it (the mount weighed 67lb and the guns 40lb), the mount would have obstructed the commander's field of view and the handlebars would have been a considerable nuisance inside the turret even when folded away. Moreover, the suggested complement of 20 drums of ammunition would have been a problem to stow in an already crowded tank. There certainly would not have been room for the guns and PLM mount itself.

There do not seem to be any references to the guns being used in action, or photographs of them mounted on a Churchill.

The turret in the photograph to the left shows a PLM mounting plate, covered by a small mushroom-shaped dome very like the one used for the ventilation fans. This may have been a way of keeping rain and shell fragments out, or a later blanking plug.

The Churchill's weapons

The 2-pounder (40mm) gun as the early Churchill's main armament was inadequate by the time of the Tunisian campaign in 1943 and it was replaced first by the 6-pounder and then by the 75mm gun. There has been much written and speculated about the merits and demerits of these weapons. An official and contemporaneous account, in the Prime Minister's response of 2 August 1944 quoted above, countered a suggestion that the 75mm gun was all that was needed, and that the 6-pounder was obsolete:

'The assumption [made by the Committee] that no Churchills are required with 6-pdr guns is incorrect. In fact Churchills with 6-pdr guns have been effective in Italy and Normandy, and General Montgomery has asked that one-quarter of his Churchill tanks should be equipped with 6-pdr guns, this being in line with existing General Staff policy. He reports in particular that the 6-pdr with special H.V. [High Velocity] armour piercing ammunition [the DSAP round explained on page 42] is a very good weapon and will penetrate the Panther anywhere, except frontally on the sloping plate.

'The Committee claim that, owing to its ineffective H.E. shell, [the 6-pounder gun] has been considered inadequate as a general tank gun since December 1942, and they ask why there has been so long a delay in replacing it by the 75mm gun. The fact is, however, that the 6-pdr gun, with its superior A.P. performance as compared with the 75mm, remains a standard General Staff requirement for an appreciable proportion of tanks.

'The official General Staff policy on tank armament during 1942 required the tank gun to be a first-class anti-tank weapon and, secondarily, as effective as possible against personnel and lorries. Battle experience at the end of 1942 and evolution of tank tactics caused a change and on the 3rd January, 1943, the General Staff laid down that the main armament of the greater proportion of medium tanks should be an effective H.E. weapon, and, secondarily, as effective as possible against enemy armour. The rest

were to have an armour piercing weapon of high performance.

'The Ministry of Supply proceeded without delay to develop the 75mm gun. It is the case that there were difficulties and disappointments. The new gun began to be issued to troops within 9 months of the decision to adopt it (i.e. in October 1943). However, after its introduction into the Service, weaknesses in the semi-automatic gear became apparent and to overcome these important modifications had to be introduced. … Trials were carried out on the normal scale which has proved adequate for other guns and this is one of the cases where weaknesses were revealed in service, which were not shown up in acceptance trials. The necessary modifications were rapidly introduced and all tanks mounting the 75mm guns are now completely modified to the approved standard.

'The principal points brought out by General Montgomery's Memorandum, the War Office Comments thereon and the latest operational reports are as follows:

'The 17-pdr and the 6-pdr with Sabot ammunition are extremely effective. Although General Montgomery draws attention to the weakness of the 75mm as an armour piercing weapon, its H.E. shell is good and as an all-purpose weapon the gun is still popular, further it proved of great value in the assault phase both in normal tanks and in D.D. Sherman and Flail tanks.'

ABOVE The mount and 2-pounder gun for a Churchill Mark I or II. Nearest to the camera is the shoulder piece that the gunner, sitting on the left, would use to elevate the gun. The curved metal item behind the gun is the cartridge catcher – as the gun recoils it ejects the shell case against this catcher, and it then drops into a canvas bag below. The bag is not fitted in this photograph. (Tank Museum)

RIGHT This
photograph shows
the contrasting sizes
of German and British
tank ammunition.
From the left: three
German rounds – the
88mm (HE), 75mm
(AP) and 50mm (AP),
compared with the
British 6-pounder solid
shot (and just in front
of it a 2-pounder shot),
6-pounder APCBC,
6-pounder Discarding
Sabot, 75mm solid
shot and 75mm AP
Capped. *(Author's
collection)*

ABOVE RIGHT In
this view can be
seen the contrast
in size between the
6-pounder HE round
(left) and the 75mm
HE round beside it. On
the right is a 75mm
smoke round (note the
flat head), and furthest
to the right is a 2in
mortar smoke round.
(Author's collection)

The Churchill's main guns

The armour-piercing ammunition for the
6-pounder was initially a simple solid shot,
followed by an improved Armour Piercing,
Capped (APC) round and then an Armour
Piercing, Capped, Ballistic Capped (APCBC)
round. This series of evolutions improved the
performance of the basic solid shot, initially by
adding a blunt cap on the tip of the round that
helped it to penetrate armour at an angle, and
then by fitting a thin metal ballistic shield to
improve its aerodynamics. Towards the end of
the war, a Discarding Sabot Armour Piercing
(DSAP) round was developed, which gave
much improved armour-piercing capability.

There was also a high-explosive (HE) round
for the 6-pounder which, due to its relatively
small size, was not particularly effective.

The Churchill Mark V and the small number
of Mark VIII tanks were classed as close-support
(CS) tanks and had a 95mm howitzer as their main
gun, with a variety of types of ammunition: HE,
smoke and armour-penetrating (shaped charge).
This gun was a more effective replacement for the
3 inch howitzer of the early tanks and mainly fired
HE and smoke rounds, which were significantly
more effective than the smaller 75mm equivalents.
The HE rounds had a range of at least 4,800yd
and the smoke rounds 1,800yd.

None of the early Churchill guns had elevation
gear; the guns were balanced – once a round
was in the chamber – so that they could be easily
moved up or down by the gunner putting his
right shoulder into a horseshoe-shaped bracket
and using this to control the elevation of the
gun. This had the mainly theoretical advantage
of allowing aiming on the move, but it did make
for quick aiming and the gun could be locked in
position once in the correct elevation. Geared
elevation was introduced for 6-pounder guns and
also for the 75mm and 95mm weapons. One

LEFT This is an unusual scene. The Churchill is a
Mark III and it looks very complete, but a 75mm
gun is being hoisted towards it with the breech
ring attached. This provides an opportunity to
appreciate the size of the gun in relation to the
rest of the tank. Normal practice is to fit the barrel
through the opening in the rear of the turret (just
visible above the track guard) and then to fit the
breech ring once the barrel is in the mantlet (as it
will not fit through the hole). The track on this tank
is the old heavy cast type. *(Tank Museum)*

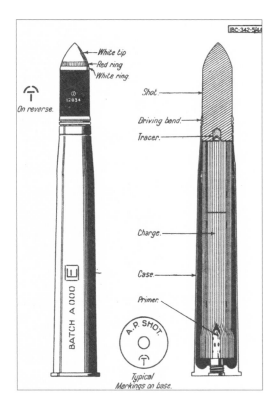

White tip
Red ring
White ring

On reverse.

Shot.

Driving band.

Tracer.

Charge.

Case.

Primer.

BATCH A 000

A.P. SHOT.

Typical Markings on base.

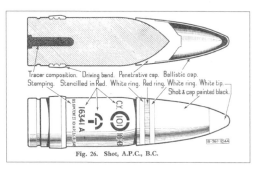

Tracer composition. Driving band. Cap. Shot & cap painted black.

Stamping. Stencilled in Red. White ring. Red ring. White ring.
White tip.

APC 6PR 7CWT XIV T
C.Y. (10) 5/43
16341 A

Fig. 25. Shot, A.P.C., Mark 14T.

Tracer composition. Driving band. Penetrative cap. Ballistic cap.
Stamping. Stencilled in Red. White ring. Red ring. White ring. White tip.
Shot & cap painted black.

6S 6PR 7CWT XIV APC B.C. T
C.Y. (10) 10/43
16341 A

Fig. 26. Shot, A.P.C., B.C.

LEFT The first modification to the 6-pounder AP round was the addition of a cap to make the round more effective when hitting armour at an angle. *(Author's collection)*

LEFT The next evolutionary step in the 6-pounder was the addition not only of the cap to improve effectiveness on impact, but also a streamlining 'ballistic' cap to improve aerodynamics. *(Author's collection)*

ABOVE A 6-pounder shot, the first of the 6-pounder anti-tank rounds. *(Author's collection)*

BELOW The 6-pounder HE rounds. Note the comparatively small amount of explosive in the shell, which itself is only 57mm in diameter. *(Author's collection)*

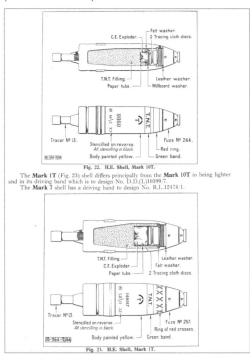

Felt washer.
C.E. Exploder. 2 Tracing cloth discs.

T.N.T. Filling. Leather washer.
Paper tube. Millboard washer.

Tracer Nº 13. Fuze Nº 244.
Stencilled on reverse. Red ring.
All stencilling in black. Green band.
Body painted yellow.

Fig. 22. H.E. Shell, Mark 10T.

The **Mark 1T** (Fig. 23) shell differs principally from the **Mark 10T** in being lighter and in its driving band which is to design No. D.D.(L)10399/7.
The **Mark 7** shell has a driving band to design No. R.L.12474/1.

T.N.T. Filling. Leather washer.
C.E. Exploder. Felt washer.
Paper tube. 2 Tracing cloth discs.

Tracer Nº 13. Fuze Nº 257.
Stencilled on reverse. Ring of red crosses.
All stencilling in black. Green band.
Body painted yellow.

Fig. 23. H.E. Shell, Mark 1T.

BELOW This is an extract from captured German research on AP projectiles. The set of high-speed photographs shows (indistinctly) a capped AP shot hitting armour plate at an angle. The sequence should be viewed starting at '3' on the top left, working down each set of three pictures, rather than across. Without the cap, the shot would have shattered. The cap provides a number of advantages: first, it destroys the hardened face of the armour; second, it deforms and resists the shot's tendency to deflect upwards; and third, it protects the point of the shot so that it can penetrate the armour at a suitable angle. This can be seen in the final picture, '20'. *(Author's collection)*

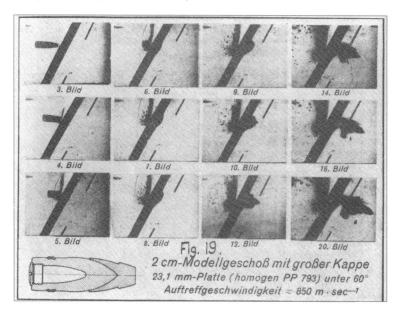

3. Bild   6. Bild   9. Bild   14. Bild
4. Bild   7. Bild   10. Bild   16. Bild
5. Bild   8. Bild   12. Bild   20. Bild

Fig. 19.
2 cm-Modellgeschoß mit großer Kappe
23,1 mm-Platte (homogen PP 793) unter 60°
Auftreffgeschwindigkeit ≈ 850 m · sec⁻¹

Auftreffgeschwindigkeit $\approx 850\ \mathrm{m \cdot sec^{-1}}$

THE CHURCHILL STORY

report on Churchill Mark III firing trials at Lulworth in February 1942 has enlightening comments that show how difficult the early geared elevation systems were for the gunner, particularly when the tank was moving: 'It was found that the absence of a shoulder piece was most noticeable and consequently gunners found it necessary to cling to the traverse spade grip and elevating gear, which is not conducive to good shooting. A small pad was attached to the gun for the gunner's right shoulder. This greatly improved the gunner's firing position.' Also, 'The blast from the 6-Pr gun when firing the first and second rounds blew dust etc. through the mounting into

THE NEW ARMOUR-PIERCING ROUND – DISCARDING SABOT ARMOUR PIERCING

RIGHT The Discarding Sabot, Armour Piercing (DSAP) round assembled. *(Author's collection)*

FAR RIGHT The DSAP round in the barrel and upon leaving it. The moment when the Sabots separate from the projectile is a notorious source of instability. This remained so until more recent developments led to an understanding of what happens as the propellant gases rush past the projectile and the separating petals. Post-war APDS rounds allow the sabot petals to lift clear of the core rather than to pivot out of the way. Also, with higher muzzle velocities, the core is now more like a dart and is termed a 'long rod penetrator'. The 6-pounder DSAP round was progressively less accurate after 500yds because of the instability that followed separation. *(Author's collection)*

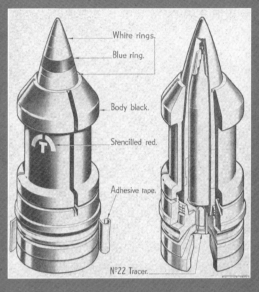

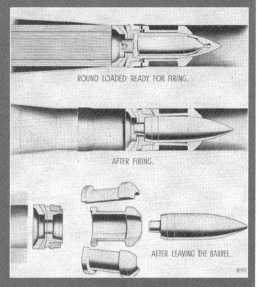

The first real leap in armour-piercing technology using a solid shot came with the Discarding Sabot Armour Piercing round. This type of ammunition is still in use today, now referred to as Armour Piercing Discarding Sabot, or APDS. The only real difference between the modern version and the old is the fact that the new versions have long projectiles with stabilising fins, rather like darts – thus known as Armour Piercing, Fin Stabilised, Discarding Sabot.

The DSAP round was, for its time – early 1944 – a secret and very successful innovation. Made from ten separate components, it has two major elements that can be seen in the accompanying photographs. The first is an outer holder made up of three segments, each termed a 'sabot', which forms the same diameter as the gun calibre, and the second, a small tungsten projectile held within the sabots. In the barrel, the composite projectile is relatively light for its diameter, so it accelerates to a high velocity. On leaving the barrel, the

sabot petals detach, allowing the projectile – aerodynamically as well as ballistically efficient due to its small cross-sectional area – to travel on to the target.

During the war only two calibres of DSAP ammunition appear to have been issued: 57mm for the 6-pounder and 76.2mm for the 17-pounder.

Published figures for the early 6-pounder DSAP rounds vary, but some quote a doubling of penetration compared to previous ammunition at 500yd. Accuracy was not good at longer ranges due to the instability caused by the detachment of the sabot petals on leaving the muzzle.

To draw further from the Cabinet Office report quoted above, which clearly shows the high esteem in which the new ammunition was held: 'The new secret 6-pdr special shot ... has given the penetration expected of it [and] appears to have had an even greater lethal effect after penetration than was foreseen.'

the gunner's face stinging him badly.' Again, 'Flash is very blinding and observation of shot strike at ranges up to about 800 yards becomes very difficult and generally impossible. Over 900 yards no difficulty is experienced [because the flash would have subsided] provided the trace functions. Quite a proportion of the ammunition used to date has failed to trace or the trace has died out under 900 yards.'

The 75mm became the standard weapon for the Mark VI and VII tanks once the requirement for HE capability became paramount. This British 75mm gun used essentially the same breech ring and block as the 6-pounder, with different extractor fingers. It would fit in the same mantlet and the ammunition stowage was basically interchangeable, although for converted Mark III and IV tanks stowage diagrams suggest that a slightly reduced number of rounds was carried. This meant that the change from 6-pounder to 75mm could be carried out in the field, as well as the factory.

The barrels were, of course, different. The 75mm gun was able to fire American 75mm ammunition, which was available in quantity. The design of this ammunition owed its origins to a First World War French gun and it was not well suited to an AP role owing to its low velocity in relation to the calibre. However, the HE and smoke rounds were effective and the ballistic-capped AP round was reasonably useful, though inferior to the 6-pounder equivalent.

The muzzle velocity of the guns is set out below and tells some, but not all of the story of how effective they were in an armour-piercing role. The actual velocity varies depending on the type of round being fired (smoke, HE, AP and so on) as well as on the amount of propellant being used. In the case of the 6-pounder, it also increased with the introduction of the longer barrel for the gun. In general, higher muzzle velocity matters most when firing solid armour-piercing rounds.

A comparison of muzzle velocity of the different main guns in the Churchill is given here:

2-pounder (40mm calibre), 2,600ft/sec;

6-pounder (57mm calibre), 2,600 to 2,950ft/sec;

3in howitzer, 600ft/sec;

75mm, 2,030ft/sec for the armour-piercing, capped M61, 1,980ft/sec for the HE M48 round (with normal charge) and 850ft/sec for

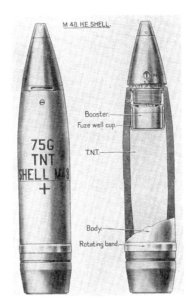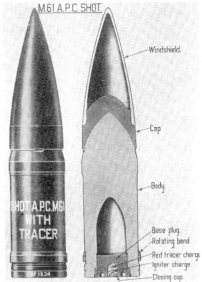

the smoke, base emission Mark II round, which being flat headed, was described as having an 'erratic flight'; and

95mm howitzer, 1,050ft/sec for HE, 655ft/sec for smoke and 1,700ft/sec for hollow charge.

Clearly, the amount and type of ammunition carried is very important as well.

The Churchill Marks I and II carried 150 rounds of 2-pounder ammunition and in the Mark I, 58 rounds of 3in howitzer ammunition. Ammunition for the Besa machine gun was 22 boxes in the Mark I and 44 boxes in the Mark II. Each box held 225 rounds of 7.92mm ammunition. Both of these tanks had 25 bombs for the 2in bomb thrower, 42 (20-round) box magazines for the Thompson sub-machine gun, 6 hand grenades and 20 flare cartridges (green, red and illuminating).

In the Mark III tank with a 6-pounder, the stowage was 84 rounds of 6-pounder ammunition and 35 boxes of Besa ammunition. Additionally it carried up to 600 rounds (6 magazines of 100 rounds each) of Bren gun ammunition, 42 20-round box magazines for the Thompsons, 6 hand grenades and the same 20 flare cartridges.

The Mark V tank carried 52 rounds of 95mm howitzer ammunition (said in the manual to be split into 28 HE, 18 smoke and 6 hollow charge), as well as 29 boxes of Besa ammunition, 600 rounds of Bren ammunition, 840 rounds of Thompson sub-machine gun ammunition or 512 rounds of Sten-gun ammunition, 30 2in mortar rounds and 12 hand grenades as well as 18 flare cartridges.

ABOVE The HE and AP rounds for the 75mm gun. Note that the AP round is capped and has a ballistic cap, but nonetheless it was only referred to as 'armour piercing, capped'. The HE round had different fuses: on the round illustrated, the setting 'DELAY' and 'SQ' can be seen at the base of the cap. 'SQ' stood for 'super quick' and would allow the shell to detonate on hitting branches or undergrowth. This had an effect like an air-burst. However, it was often not the result that was needed, and the delay setting would allow the shell to continue to its target before exploding. (Author's collection)

RIGHT The cartridge catcher behind the 6-pounder gun in a Mark IV. Ejected shell cases hit the back plate and fall into the canvas bag below. The catcher folds forward on to the top of the breech ring. When in position, as in the photograph, it prevents anyone moving behind the gun. This had tragic consequences for one 9th Royal Tank Regiment (9 RTR) crew when the tank caught fire and the commander's hatches jammed. Although the loader was able to get out, the commander and gunner could not get past the gun to escape through the other turret hatch, and died. *(Author's collection)*

By the time of the Mark VI Churchill, which was essentially an upgraded Mark IV, the tank is in transition from using the old, unprotected ammunition stowage to armoured bins similar to those used in the Mark VII tanks.

For example, the manual for gunnery in the Mark VI Churchill sets out that it was originally supposed to carry 87 rounds of 75mm main gun ammunition, broken down as 28 AP (APC M61), 23 HE (M48) and 36 smoke. After the stowage was changed to use armoured bins, it became 84 rounds stored as follows: 14

stowed nose down in the turret floor basket and 30 rounds stored horizontally (in six tiers of five rounds) in each of the two armoured bins in the hull. Additionally, 10 rounds of smoke ammunition were stowed in two bins accessible only to the driver and front gunner. In the modified Mark VI and the Mark VII, there were also 29 boxes of Besa ammunition, 6 drum magazines of 100 rounds for the Bren gun, 36 box magazines for the Thompsons and 20 rounds of 2in smoke bombs (down from 30 in the more exposed open trays used in the Marks III and IV). Interestingly, in the new stowage for the 2in bomb-thrower, the highly volatile white phosphorous rounds were to be kept in the centre section of the closed stowage bin, which was a tray that held 10 rounds, so that they could be 'jettisoned easily in the event of fire'.

These tanks carried nine hand grenades, an increase from the previously held stock of six.

The Mark VII had two of the same type of armoured bins as the Mark VI, storing 30 rounds in each plus the additional 5 smoke rounds on each side behind the driver and front gunner. However, the bin on the turret floor was lower and smaller and took 15 rounds, which gave the tank a total of 85 rounds for the main gun. Other stowage was the same as for the upgraded Mark VI.

The standard guidance on mixture of ammunition was given: 'The proportion of AP to HE will vary according to the tactical situation. At the beginning of a battle, it is suggested that the tanks carry 50 per cent AP and 50 per cent HE, and that if the situation becomes static the amount of HE should be increased.'

Finally, there was the question of what happened to the empty shell cases once the main gun had fired. Once the gun discharged, it recoiled and the breech block dropped. The extractor fingers flicked the shell case out and it hit a 'catcher' behind the gun and was deposited into an open-bottomed canvas bag.

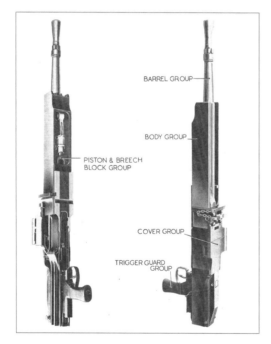

RIGHT The Besa 7.92mm machine gun seen from beneath and the side. *(Author's collection)*

BARREL GROUP

BODY GROUP

PISTON & BREECH BLOCK GROUP

COVER GROUP

TRIGGER GUARD GROUP

RIGHT Some examples of the bases of Besa ammunition showing the coloured annuli that indicate the type of bullet (there were no markings on the tips of the bullets), and markings. *(Author's collection)*

The Besa 7.92mm machine gun

Approximate weight: 48lb.
Overall length: 3ft 7.5in.
Rate of fire: varies with mark, but generally – high, 750/850 rounds per minute; low, 450/550 rounds per minute.
Muzzle velocity: 2,700ft/sec.

The Besa was unusual in British Army service not least because its calibre, that of the standard German machine guns and rifles, was 7.92mm rather than the normal .303in of the Bren, Vickers and Lee Enfield guns. It was manufactured by BSA based on a Czech design. The Besa was almost exclusively used as a weapon in armoured vehicles and had neither a ground mounting nor its own sights. As can be seen, it had a relatively slow rate of fire and, because of this and its air-cooled heavy barrel assembly, could be used for sustained firing without needing a barrel change. This was just as well, since changing the barrel involved removing the gun from its mounting.

It fired AP, incendiary, tracer and ball rounds.

The 2in bomb-thrower

See box on page 22 for details of these.

The 4in smoke emitters

See box on page 23 for details of these.

Crew personal weapons

Each tank also carried crew weapons comprising one Bren .303in light machine gun (mainly intended for anti-aircraft use but rarely used in that role), two Sten (9mm) or Thompson (.45in ACP) sub-machine guns. Veterans of 9 RTR recall having their Thompson sub-machine guns taken away to be given to the infantry and getting Sten guns instead. Certainly many contemporary photographs show the Sten as being the crew weapon.

Crew members generally armed themselves with pistols or revolvers. The tanks also carried a 1in flare gun and a stock of different coloured flares (red, green and illuminating/white). The Bren gun in the tank normally used a circular-drum 100-round magazine, rather than the traditional curved one. This magazine required a special adaptor to be fitted to the gun and had the disadvantage that the 'peep' sight that worked with the standard curved magazine on the gun could not be used, as the circular magazine sat across the line of the sight. The drum magazine was hand loaded, using a special tool to turn the spring inside to allow the rounds to be fed in. When fired, it started at a low rate of fire, and then picked up speed.

Hand grenades

Hand grenades were carried on board – usually either explosive or smoke. The stowage for grenades was initially an open rack for six grenades. This proved to be particularly vulnerable and was replaced by three strong steel boxes each holding three grenades. The grenades carried tended to be No 36 (explosive, commonly known as the Mills bomb), No 80 (white phosphorous, bursting smoke) or No 81 (duplex, combining a white phosphorous bursting action with smoke generation).

THE ROTA TRAILER

One answer to the lack of stowage inside the Churchill was this Rota Trailer, to be towed behind the tank carrying petrol in the wheels and ammunition and other stores in the body, as can be seen from the stowage diagrams. It was reputedly hated by the crew for being difficult to tow, particularly when reversing, as well as having a tendency to leak, which made it a danger to the crew if they were under fire. (Tank Museum)

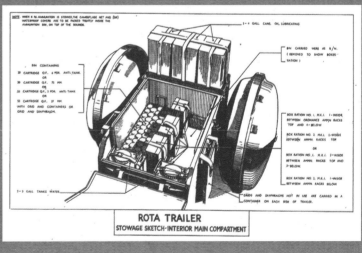

ROTA TRAILER
STOWAGE SKETCH-INTERIOR MAIN COMPARTMENT

Chapter Two

The 'Funnies' and specialised armour

The Churchill was not just a gun tank; it evolved into countless specialist adaptations to help the soldiers fight their battles. Whether in support of beach landings or in the slog across well-defended terrain, the Churchill became like a Swiss Army Knife – a tank with many adaptations and attachments.

OPPOSITE A rare opportunity to see what a Churchill 'Toad' mine-clearing flail looks like in use. The flail is being used on straw to give a sense of the effect of the action. The flail chains have been shortened for this demonstration. *(Author's collection)*

47

THE 'FUNNIES' AND SPECIALISED ARMOUR

ABOVE A Churchill Mark IV prepared for wading. Air intake trunks have been fitted to the sides, and one at the back to allow air out of the engine compartment. Note how the exhausts have also been extended. What this photograph does not show is the extensive waterproofing of other openings lower in the hull. For those who study the different castings of Mark IV/V/VI turrets, this one has the cover for the ventilation fan moved a little to the right of the earlier types. The cover is protected by a lip raised around the hole in the turret to stop bullet splash, which was a risk with the previous version. Later turrets had the fan in an even more central position. (Tank Museum)

and had side, or 'pannier', doors allowing easy access from outside so that stores, ammunition and other items could be moved in and out of the tank from ground level. Of course, having pannier doors was a handicap when it came to wading through deep water, and the Churchill was not chosen, as the Sherman was, to be an amphibious tank to 'swim' from landing craft. The Churchill was only capable of deep wading after lengthy preparation involving sealing the various openings around the hull. Nor was it readily adapted to clearing mines as a flail tank; again, the Sherman, which had its driving sprockets at the front of the tank, was the main choice for this role in the Second World War. The Churchill did attain the role of a flail tank after the war – see the 'Mine and obstacle clearing' section on page 55. Pretty well every other possible adaptation that could be made on a tank was put on the Churchill during its evolution.

Some of the roles in which the Churchill proved most successful were as an engineer vehicle. However, there was also a notably

One of the remarkable features of the Churchill tank was that, aside from being a very successful infantry tank, it was also adapted for use in a large number of variants dedicated to special purposes. This was probably due to two key attributes: it was relatively spacious inside;

MAJOR GENERAL PERCY C.S. HOBART

The story of Churchill adaptations has to be told alongside a tribute to this remarkable soldier. He served in the First World War, gaining both the Military Cross and the DSO as well as being mentioned in dispatches five times. He and other visionaries had argued, between the wars, that the British Army should have a separate tank force, but while the German Army took these ideas to heart, the British establishment did not. After an unpromising start to the Second World War, in which 'Hobo', as he was known, was first given a command, then dismissed and placed in the Reserve, his skills were recognised by Churchill and Montgomery and in October 1942 he was given command of the 79th Armoured Division. In this role he had the task of leading the 79th in the development of tactics and organisation for what were to become known as 'Hobart's Funnies'

– specialised vehicles to help with the assault on Europe. Of course, many of the Funnies were Churchill-based. These tanks made a very considerable difference to the British landings in Normandy and to other operations all the way to the crossing of the Rhine in Operation Varsity in March 1945. They were retained within the 79th Armoured Division and released for use in small groups by other units as needed.

RIGHT The bull emblem of the 79th Armoured Division. (Author's collection).

ARMOURED DIVISION

successful version of the Mark VII as a flame-thrower – the Crocodile.

Space does not permit a full review of these variants here; they are set out in some of the books listed in the Bibliography. However, it is important to note at least a selection of these tanks and their roles in any account of the Churchill.

Engineer and related Churchill roles

Armoured Vehicle Royal Engineers (AVRE)

The AVRE tank was developed as a direct consequence of the Dieppe landings, and was largely the brainchild of a Canadian officer, Lieutenant John James Denovan of the Royal Canadian Engineers, working in the Special Devices Branch of the Department of Tank Design (DTD) in England. The Assistant Director of the Special Devices Branch of DTD was

ABOVE A well-used Churchill AVRE of the 79th Armoured Division (the bull emblem), 42nd Assault Regiment (denoted by the number '1235' on the hull). Note the raised periscope holders for the driver and front gunner to see over the track guards. On this tank, unusually, the ultraviolet lamp is on the front gunner's side, rather than the driver's. There is a tank crew helmet hanging in front of the blade vane sight on the right of the turret roof. *(Tank Museum)*

LEFT This is a photograph looking down on the Petard Mortar, in which the fins of the 'Flying Dustbin' round are just visible inside the base. The vehicle is being waterproofed before an amphibious landing and the front gunner's sliding hatch has been covered with the canvas and waterproofing material. The spring to the right of the Petard helps it to 'break' like the barrel of a shotgun, so that the front gunner can fit a new round. *(Tank Museum)*

BELOW LEFT A close-up of the Petard in a Mark IV AVRE. The Besa muzzle can be seen in its normal position. Note the pipe running past it to the Petard, carrying the blast from a cartridge set off inside the turret to ignite the propellant in the Petard round. The handle to break open the Petard is below the barrel. *(Tank Museum)*

BELOW An enthusiastically driven Mark IV AVRE, *Liberator*, with appliqué armour. *(Tank Museum)*

LEFT A Churchill Mark IV AVRE which has been waterproofed – note the canvas over the Besa mount and around the mantlet, as well as over the round already loaded in the Petard. The tank has also been modified for deep wading, with tall trunking on top of the air inlet louvre on the left of the photograph. This enables it to wade without water being drawn into the engine. The canvas incorporated a detonating cable that was set off from inside the tank to blow off the water-proofing once on dry land. *(Tank Museum)*

412.M.1.
INTERNAL VIEW OF MOUNTING.

412.M.2.
RE-COCKING DEVICE AND FIRING LEVER

ABOVE Inside the turret of an AVRE. In the first picture the view is through the loader/operator's hatch; and in the second, through the commander's cupola. The large cylinder with a circular plate bolted at the back is the rear of the Petard, which contains a very large spring. In the photograph on the right taken through the commander's hatch, is the circular 'magazine', which holds the cartridges that set off the Petard round. The blast from the cartridge goes down a pipe to the back of the round and sets off the propellant. To the left of the Petard fitting, the Besa machine gun and the gunner's telescope can be seen. *(Tank Museum)*

RIGHT This photograph gives some idea of what a Spigot Mortar comprises. Note the spigot at the front, over which the round slides. The giant spring in the foreground holds the recoil. *(Tank Museum)*

Lieutenant Colonel George Reeves, who had served as an observer on the Dieppe raid.

After Dieppe Colonel Reeves wrote a report about the problems faced in the operation, concluding that the most difficult job had been that given to the Royal Engineers. Reeves must have been aware of how serious the casualties and fatalities were among the Engineers at Dieppe as they endeavoured, unshielded, to destroy obstacles under intense enemy fire in order to enable the tanks to move off the beach.

Denovan believed that in combat, Engineers needed protection to work at the front line and that a modified tank was the best way to provide this.

He decided to work on adapting the Churchill to this role, and in October 1942 one was stripped out and the interior reconstituted as a mock-up of stowage arrangements suitable for use by Engineers who needed to carry explosive charges, fuses and other equipment. What made this variant of the tank particularly ingenious was the decision to go further and make the tank not just a form of armoured personnel carrier, but a vehicle capable of projecting an explosive charge on to its target without making the crew dismount. To be useful, this explosive charge had to be very large and heavy, and this put it beyond the capability of any then-available gun that might fit inside a tank turret and be able to recoil within it.

Denovan enlisted the help of Colonel Stewart Blacker, inventor of the Projector, Infantry, Anti-Tank (PIAT) as well as the Home Guard weapon, the Blacker Bombard. Both of these weapons were, in effect, spigot mortars.

A spigot mortar is a clever variation on the normal use of a barrel to aim a projectile. The

RIGHT A dramatic image of a Mark IV AVRE, T172622B, at the moment of releasing its Fascine bundle from the top of a high obstacle.
(Author's collection)

principle relies upon a solid rod (hence the term 'spigot') over which the projectile slides. The projectile has a hollow tube running back to fins at the rear and this tube fits snugly over the spigot. In the front of the tube is the propellant. When set off, the propellant expands and forces the projectile along the spigot, which serves to guide and stabilise it, and sends it on its way. The weapon is relatively simple to make and well suited to firing over limited ranges.

Blacker designed a tank-mounted spigot mortar, which was christened the 'petard', capable of firing a highly effective projectile of some 28lb over approximately 100yd; it came to be nicknamed the 'flying dustbin'.

On 14 January 1943, following a demonstration of the prototype, the concept received War Office approval and was put into production using conversions of Mark III and Mark IV Churchills. One particular cleverness of the petard lay in the fact that it could be fitted to the existing mantlet as a direct replacement for a 6-pounder or 75mm gun. Its disadvantage lay in the comparatively short range of the weapon and the fact that loading it exposed the front gunner, who had to open a sliding overhead hatch (in place of the normal opening flaps) and break the barrel to insert the heavy round and close it again. This was not an operation that would be wise to carry out under fire.

The AVRE tanks carried a series of additional items in service, notably fascines – giant bundles of wood that could be dropped into obstacles – and also a bridge (the 'small box girder' or 'assault' bridge) that could be lowered from the front of the hull and would carry a 40-ton load.

Another vital role to which AVREs were adapted was that of laying a matting to enable other tanks to cross soft beach areas. This was done by carrying a giant 'bobbin' on two arms at the front of the tank for mounting a roll of canvas and/or wooden stakes. This was then driven over by the tank itself so as to form a track on which other vehicles could follow.

Various adaptations were provided for the Churchill and some 24 of them were used in the D-Day landings.

Post-war Churchill AVRE

Such was the success of the wartime AVRE concept that, between 1947 and the early

ABOVE Churchill Mark IV AVRE, *Desirée*, photographed after the war with a bridge attached. These bridges were very cumbersome and highly conspicuous on the battlefield.
(Tank Museum)

LEFT A Churchill Mark IV AVRE laying matting. *(Tank Museum)*

ABOVE A post-war Mark VII AVRE conversion named *Mars*, carrying a fascine bundle and fitted with the 6.5in breech-loading gun. Note the loudspeaker on the turret side, just ahead of the cable reel. These tanks carried two such speakers, facing to the rear, which allowed the commander to be heard by troops following behind. *(Tank Museum)*

1950s, some 88 Mark VII or VIII Churchill tanks were converted into an improved version of the Churchill AVRE carrying a breech-loading 6.5in demolition gun. While the original acronym had stood for Armoured Vehicle Royal Engineers,

LEFT Another post-war AVRE conversion, this one called *Mons II*, with the later version of the 6.5in gun, fitted with a fume extractor. This tank also has a dozer blade. *(Tank Museum)*

this new, post-war, Churchill AVRE was designated as Assault Vehicle Royal Engineers.

The gun fired a type of shell known as High Explosive Squash Head (HESH) which, as the name implies, first squashes against the target on impact, and after a momentary delay to allow the squashing process to occur, uses a fuse in the base to detonate, sending a shockwave through the target. The round had no shell case in the normal sense, but a perforated base containing the propellant and attached to the projectile.

The primary role of the new Churchill AVRE was in the use of its demolition gun, but the tank also carried demolition charges and detonators, which a member of the crew could place by hand. It was also intended to fulfil other functions including that of bulldozer, mobile crane, bridging, tractor and fascine layer.

Despite the fact that work on these tanks began in 1947, it was 1954 before they officially entered service, so they were probably some of the last Churchills to serve with the British

Army. So far as is known, they were never used in action. In 1955 design work began on their replacement, the excellent, versatile and very long-serving Centurion AVRE.

ARK

During the war there was a pressing need to enable tanks to advance across the many defensive anti-tank obstacles in the form of ditches excavated by the Germans in Italy and north-west Europe, as well as smaller streams and large shell craters. To do this, Churchills

ABOVE An early pattern of Churchill ARK. *(Tank Museum)*

BELOW LEFT A Churchill Ark deployed to enable Sherman T211770, *Spiteful*, to climb a steep bank. Curiously, this ARK has had its air intake louvre removed on the side facing the camera. *(Tank Museum)*

BELOW A pair of post-war linked ARK tanks. They are coupled together (see the fitting between them). The tank on the left is a pre-Mark VII (note the rectangular opening for the driver's vision port, as opposed to the round one on the Marks VII and VIII). The earlier tank has had the Besa mount removed from the front, whereas the tank on the right, a converted Mark VII or VIII, has retained its hull Besa mounting. *(Tank Museum)*

were adapted by removing their turrets and providing fixed decking over their tracks so that they could drive into the obstacle, and other tanks could then literally drive over them. Various patterns of ARK (armoured ramp carrier) emerged during and after the war, but in their earliest form, other tanks simply drove over the tracks of the turretless ARK. In its final, post-war, form, the ARK became a 'Twin ARK', which comprised two ARKs that could link together to enable the Conqueror and Centurion tanks to cross obstacles.

Bridge layer

Although the ARK was very useful for crossing shorter, usually dry, obstacles, there was also a need to cross rivers and longer gaps. For this, the Churchill was turned into a very successful turretless bridge layer capable of deploying a bridge without exposing the crew unduly.

Armoured Recovery Vehicle (ARV)

This was an initially turretless variant of the Churchill tank, created in order to enable the fitters of the Royal Electrical and Mechanical

LEFT An early ARV
uses its jib to lift a final
drive assembly onto
a Churchill Mark IV.
(Tank Museum)

Engineers (REME), to reach damaged Churchills
and other vehicles in comparative safety, and to
tow them out of trouble, or carry out heavy repairs
beyond the scope of their crews. This vehicle,
with increasing levels of sophistication, allowed
REME fitters to lift engines and gearboxes, change
suspension units after mine damage and weld
and repair battle damage with all of the necessary
tools and equipment on board their own vehicle.
The ARV eventually was equipped with a non-
traversing turret and dummy main gun. Such ARVs
were still in service during the Korean War.

Mine and obstacle clearing

In the war many variations of mine roller and
plough were tried using the Churchill as a basis.
Additionally, a series of frames to place explosive
charges on obstacles was also tested on the
tanks. Of all the variant roles for the Churchill,
these seem to have seen the least active service,
notwithstanding the significant amount of time
and resource devoted to their development.

After the war Distington Engineering Company
in Cumberland was asked to oversee the design
of the Churchill as an unarmed flail tank with no

BELOW LEFT A
REME fitter gas-welds
a bracket using a vice
mounted on the front
of a Churchill AVRE.
(Tank Museum)

BELOW T38306R is
a very late variation
on the ARV theme.
This tank has a mock
'turret' with a mock
gun as well as a better
crane and towing gear.
(Tank Museum)

turret, a frontal height of more than 8ft and a crew
of two. Behind them there was an M120 Meteor
fuel-injected engine-devoted entirely to driving the
flail drum. The original engine compartment from
the Churchill was retained and the flail tank was
able to travel at the same speed as a Churchill
Mark VII – 12.7mph. However, when flailing,
it could only travel at a maximum of 2mph. It
weighed 56 tons with all equipment on board.

At the rear of the tank was a lane-marking
device that would fire stakes into the ground to
indicate the route that was cleared by the flail.
There was sophisticated equipment to help the
driver keep to his designated route – including
gyro-stabilised direction indication and also
provision to enable a second flail to keep in
position relative to the first one. Strangely for a

vehicle with so much advanced equipment, the flail, mounted on two giant arms, did not deploy automatically. It had to be pulled into and out of position by cables. The vehicle was nicknamed 'Toad' by the Army and was never used in active service.

Robinson and Kershaw Limited of Dunkinfield converted Mark VII/VIII hulls for the purpose and British Railways workshops at Horwich in Lancashire carried out the installation work. Some 42 flail conversions were made between 1954 and 1956.

Gun tank adaptations

Churchill 3in gun carrier

Following a requirement for a self-propelled gun that could take on German armour in the event of an invasion, a decision was taken to mount a 3in anti-aircraft gun to a modified Churchill hull with no turret. Fifty of these vehicles were made by Beyer Peacock & Company by the end of 1942 but, so far as is known, never used in action.

Crocodile flame-thrower

Another of General Hobart's 'Funnies', this tank was developed in great secrecy before the Normandy landings to provide – along with the AVRE and its petard mortar – a means of attacking the vast fortified emplacements used by the German Army in its defensive positions. The Crocodile was a standard Mark VII gun tank in which the hull Besa was removed and a flame-thrower placed in its mounting for the front gunner to use. Two pipes ran from this, through a hole in the floor, along the underside of the hull to a large armoured coupling at the rear of the tank. This was attached to an armoured trailer, weighing some 6.5 tons, which carried 400 gallons of flame-thrower fuel and five compressed nitrogen gas tanks that provided the means to force it through the muzzle of the flame-thrower to a distance of up to 150yd (80yd was more realistic in most conditions). The flammable liquid was poured by hand into two tanks in the trailer and the taps on cylinders of compressed nitrogen would be opened on the order being given to 'Pressure Up'. After this, the pressure would remain sufficient for flaming for only a limited time: the tank could fire 80 one-second bursts

from the flame-thrower. The liquid could either be ignited on firing, or be hosed on to the target unlit, before a subsequent burst set light to it. The trailer could be jettisoned from within the tank.

The Crocodile still carried its full complement of 75mm ammunition and its main gun, and could function as a gun tank. It was much feared by the Germans who would, when the Crocodile appeared on a battlefield, tend to concentrate anti-tank fire particularly on these tanks and their trailers. There is at least one documented instance of the captured crew of a Crocodile being executed immediately in revenge for the effect of their flame-throwers.

It was often sufficient for an AVRE and

ABOVE The Churchill gun carrier outside an 'Experimental Dept'. *(Author's collection)*

BELOW Drawing of the driver's compartment of the Crocodile, showing the flame-thrower projector in front of the front gunner's position. Compare with the photograph on page 59. *(Author's collection)*

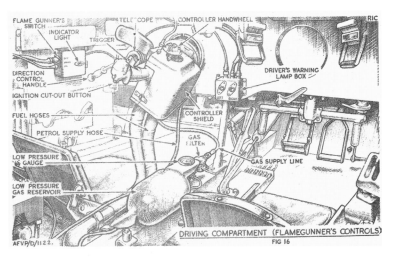

RIGHT **Diagram**
showing the range
of the flame-thrower.
(Author's collection)

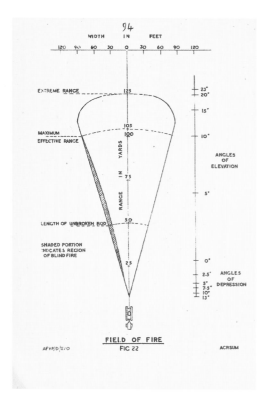

Crocodile to appear close to a stubborn fortification, and give a demonstration of their weapons, to persuade the occupants to surrender. When this did not work, a petard blast was found effective as a means to distort or blow in the frames of armoured doors within fortifications so that the burning liquid from the flame-thrower could penetrate to the parts where defenders might be sheltering.

Prior to the Crocodile, the Churchill had been used as a basis for the addition of a flame-thrower on a few early tanks to form the 'Oke' – a conversion named after Major J.M. Oke who designed it.

The Oke design was submitted in late 1941 and involved using the Ronson flame-throwing equipment that had already been installed on infantry carriers. Only limited numbers were made and the Crocodile superseded it.

Initial tests on the Crocodile flame-thrower concept were carried out on Mark IV Churchills before the design was formalised and allocated to the better-armoured Mark VII.

NA 75

This adaptation of the Churchill Mark IV is a tribute to the ingenuity and persistence of one man, Captain P.H. Morrell, who was serving with the Royal Electrical and Mechanical Engineers in North Africa. There he saw a potential benefit in taking guns and mantlets from the many damaged Shermans in Tunisia and fitting them to Churchill Mark IV tanks. This would serve two purposes: first, to enable the Sherman's 75mm high-explosive round to be used by the Churchills; second, as the Shermans had an external mantlet, to eliminate the dark shadow formed by the internal mantlet of the Churchill and provide better protection against shot damage. The Sherman mantlet was, in effect, a shield that fits in front of the turret, whereas the standard mantlet on the Churchills sat inside. In the bright North African sunshine, this was felt to give the enemy a highly visible aiming point and to be a danger to crews. Captain Morrell noted that 60% of the Churchill casualties in the Medjerde Valley

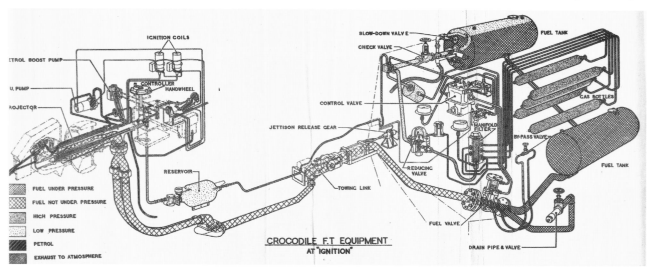

ABOVE The official caption to this photograph states: 'British flame-throwing Crocodile tanks in demonstration on German soil against a knocked out Nazi self-propelled gun. Three tanks are flaming on target.' *(Author's collection)*

BELOW A Crocodile at the entrance of Belsen Concentration Camp. Crocodiles were used to burn down the huts and buildings. *(Tank Museum)*

ABOVE Another rare photograph of the interior of a Churchill. This is a Crocodile and the flame-thrower control can clearly be seen in the front gunner's position, with its handle and trigger at the rear. The driver's seat is in the foreground. Next to the flame-thrower control on the driver's side is a box with two lights. These warned the driver if the trailer was getting close to the rear of the tank in a tight turn. *(Tank Museum)*

ABOVE To give some sense of proportion and scale, here is the trailer towed behind the Crocodile. A crew member is making adjustments on top. *(Author's collection)*

LEFT A Crocodile flaming at an unseen target. *(Author's collection)*

OPERATION WHITEHOT – THE CHURCHILL NA 75

Captain Morrell was a REME captain in Tunisia tasked with 'reduction to produce', or scrapping, of Sherman tanks that were beyond economical repair. He noted that many of the Shermans' guns were reusable, and proposed a conversion of the Churchill to accommodate them. This concept proved sufficiently interesting to those in authority for a civilian from Vauxhall – John Jack – to be dispatched to join him and for Major General W.S. Tope, who commanded REME services in the central Mediterranean theatre, to give approval. The project was classified as 'Top Secret' and given the code name 'Whitehot'. General Tope's advice to Captain Morrell (who had been commissioned only a couple of years before from warrant rank) was: 'If you can make a success of this project, which I warn you is pooh-poohed by several of the AFV experts I have referred it to, I will see that you do not lose by it. If, on the other hand, you are unable to make a job of it, and in the process you render unserviceable a tank which cost the British taxpayer a great deal of money, you can take it that your career has advanced just about as far as it is going to! Don't waste any time. I think you can do it.'

The conversion worked and this was Captain Morrell's note of its first trial:

'The trials were an unqualified success. Having got the 'feel' of the gun with a few preliminary shots, Gunnery Instructor at the RAC Training Depot at Le Khroub, Major 'Dick' Whittington, bracketed a deserted Arab village which was ranged at some 8,000–8,500 yards, and then brought down round after round of HE on top of it. He said that because the Churchill provided a much more solid firing platform than the Sherman, and did not 'rock' to the recoil of the gun to anything like the same extent, the accuracy was greater, and the range increased.'

Morrell received a letter from Major General Tope after the project finished, which included a report from 21 Tank Brigade following nearly a month's fighting between Arezzo and Florence: 'I should be glad if you would congratulate the REME concerned on doing a quick job which had been most valuable to this Brigade.'

Captain Morrell was awarded the MBE (Military) for his work.

RIGHT Major Morrell, following his promotion. He enlisted at Leeds on 29 June 1940 and rose through the ranks to be granted an emergency commission as a 2nd lieutenant on 6 February 1943. He was posted to North Africa in April of that year. He rose to become an acting lieutenant-colonel. *(Courtesy Morrell family archive)*

battles in Tunisia were hit in or around the mantlet. He was able to persuade the powers that be that the Sherman mantlet and gun would fit the Churchill turret and, after trials, this proved to be the case. Work then proceeded rapidly in the months up to June 1944 to convert some 190 to 200 Churchill Mark IV tanks to what became known as the NA (for North Africa) 75. These saw service through the Italian campaign.

ABOVE This shows the quick way to remove the Churchill's gun. Rather than dismantle the breech, the whole gun assembly is turned through 90 degrees and dropped from a raised turret, in this case the tank was T31183B. The 6-pounders were returned to Ordnance stores. *(Courtesy Morrell family archive)*

BELOW The chalk marking shows the extent of the armour that had to be cut out to convert the standard Mark IV turret into the NA 75. On some turret castings the fume extractor hole could be used for the Sherman gun's 'peritelescope' and a new hole cut to re-site the extractor fan. This was such a turret. *(Courtesy Morrell family archive)*

ABOVE An example of a finished turret. *(Courtesy Morrell family archive)*

ABOVE The complete Sherman gun and mantlet ready to install. It was necessary to rotate the gun through 180 degrees to allow the loader in a Churchill to use it. In the Sherman the loader stood on the left of the gun, which is the opposite arrangement to the Churchill. *(Courtesy Morrell family archive)*

BELOW The completed conversion. In this photograph the fact that the main gun can be elevated to a greater degree than the coaxial Browning can clearly be seen. In the background is a Churchill Mark IV with no gun. Note that the sidelights on this tank have been moved to the outside of the track guards. *(Courtesy Morrell family archive)*

ABOVE This image shows the Sherman gun assembly in the Churchill, with grease liberally applied to the breech area. Note that the stowage for the 2in bomb-thrower has been reduced to ten rounds; further rounds were stored at the back of the turret. Note also that the elevation gear has needed to be linked by a long shaft across the top of the gun, owing to the fact that the gunner in the Sherman is on the right. Barely visible on the left, above the traverse gearbox, is the Browning machine gun that was installed with the conversion. It has reached its maximum elevation and is now stopped on the top of the traverse gearbox. From this point the main gun can continue to elevate, but the Browning remains at a lower elevation. Given the useful range of the machine gun, that was not considered to be a problem. The holder for the Browning ammunition tin is underneath the mortar round stowage. *(Courtesy Morrell family archive)*

RIGHT The completed NA 75 with the photograph signed by (then) Captain Morrell saying 'Bone (where the workshops were), North Africa, May '44'. *(Courtesy Morrell family archive)*

Chapter Three

The Churchill at war

Even with its shortcomings in armament, the Churchill as an infantry tank made an indelible impression on those who served with it, as well as those it fought against. Its ability to take punishment and still fight on was as legendary as its ability to climb impossible gradients and surprise the enemy.

OPPOSITE A remarkably heavily laden Mark IV or VI, with additional water or fuel cans, tarpaulins and other items, advances with the turret traversed slightly to the right. There appears to be a plate welded onto the turret to protect the commander from sniper fire. *(Tank Museum)*

ABOVE This grainy photograph shows *Backer*, T68352, a Mark II Churchill on the beach at Dieppe. As this tank left its landing craft, it received a hit on the left side, just below the turret ring. This jammed the turret, although it did not seem to penetrate the tank. Shortly after this the tracks broke, either because of gun-fire or the stones on the beach. As the tank was facing parallel to the sea, the jammed gun could not be brought to bear on any useful target. With great bravery, the co-driver, Trooper Chapman, got out of the tank under fire and attached one end of a rope to the gun barrel and the other to the broken left track. The driver then reversed that track and slewed the turret round to face some of the buildings adjoining the landing area. The crew fired all of their ammunition and used a grenade to disable the tank before taking cover. Sadly the gunner, Trooper Provis, was then killed by a sniper while trying to take cover.

THE CALGARY REGIMENT

The Calgary Regiment traced its origin back to 1910. After training in England as the Calgary Regiment (Tank), it became the first tank regiment of the Canadian Army to engage in combat with the enemy in the Second World War. At Dieppe the regiment lost 2 officers and 10 men killed, and 157 captured. The regiment fought through Italy until April 1945 when they moved to the Reichswald Forest where they provided fire support for the crossing of the Ijsel River. The regiment then went to Holland, where they remained in action until the end of hostilities on VE Day, 8 May 1945.

The regiment was officially disbanded on 15 December 1945, and returned to reserve status.

RIGHT The cap badge of the Calgary Regiment. *(Author's collection)*

To appreciate fully the achievements of the Churchill, it is important to look at the tanks' role in battles, what they achieved and what the tank crews had to contend with. A number of the accounts that follow, on the principal fighting engagements, are drawn from the excellent books published by veterans, many now out of print but still available second-hand, which enable the stories to be told through the voices of those who were there. These are listed in the Bibliography.

The Dieppe landings

The Churchill had its first baptism of fire during the Dieppe raid on 19 August 1942, in the hands of the Canadian Calgary Regiment (14th Canadian Army Tank Regiment).

The tanks used at Dieppe were a mixture of Marks I, II and III. Three Mark I tanks appear to have had Oke flame-throwing equipment fitted in place of the howitzer.

Some 29 Churchills were launched out of the Calgary Regiment's total of 58; 2 were lost on launching in deep water, and of the 27 that landed, only 15 managed to climb the sea wall and move close to the heavily fortified promenade of the town. So far as is known, only two of the tanks showed evidence of their armour having been penetrated by enemy fire (an Allied account says none were penetrated; the German account suggests two).

Many tanks were immobilised by losing their tracks that day; although gunfire would have contributed to this, at least eight tanks had track breakages caused by the stones on the beach. The tanks were all fitted with what is referred to, even in early manuals, as 'Heavy Cast Steel Track, old type'. This track seems to have been particularly unsuited to the stony beach at Dieppe.

In many ways the raid was a tragic failure, but this in no way detracts from the bravery of the tank crews, and the infantry and engineers they were supporting on the day. The captured tanks were analysed by the German Army after the battle and their report considered them to be wanting in many respects. Some of the captured tanks were used as targets for German and, later, Russian weapons to test their effectiveness against the machines. The turret from at least one tank was used as a fixed gun emplacement nearby. The German Army's conclusion was that the Churchill was poorly armoured, noisy for the crew, primitive and under-gunned. As will be seen from the tank's subsequent history, the armour was good by the standard of British and American tanks during the war, and the Churchill was a remarkably robust tank in combat – one that the Germans and Italians came to fear and respect. What cannot be denied is that, even in 1942, it was under-gunned and remained so throughout the war.

Whatever their shortcomings in planning the raid, Dieppe provided military planners with a graphic lesson in what not to do in a beach landing, lessons that should have been known to them already. Those lessons were duly absorbed and the landings in Normandy, in which the Churchill was to play such a vital role, were an altogether different story.

However, before Normandy, the Churchill tank was to earn its fighting reputation in other battles, the next of which was in North Africa.

El Alamein and Kingforce

At the end of 1942 six Churchill Mark IIIs with 6-pounder guns were unloaded in Egypt to take part in the Second Battle of Alamein. This was largely an evaluation exercise for the tanks, referred to as 'operational experience'. The unit taking the tanks into battle, known as the Special Tank Squadron, comprised 6 officers and 52 other ranks commanded by Major King, MC, of the Royal Gloucestershire Hussars and as a consequence was known as 'Kingforce'. It fought as part of 1st Armoured Division and three of the tanks saw action on 27 October in support of the Queen's Bays. Major King's tank fired 45 rounds of 6-pounder ammunition and, surprisingly, only two belts of Besa ammunition. His tank was hit six times, four of those by 50mm and twice with 75mm projectiles. The armour was not penetrated. Of the other two tanks, one commanded by Corporal Kelly had to retire almost immediately when its 6-pounder gun recoiled but would not spring back again (a not uncommon problem with the early 6-pounder gun in Churchills). The third tank, commanded by 2nd Lieutenant Allan Appleby, advanced into heavy fire, before reversing back

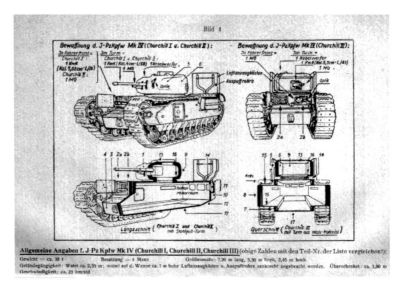

ABOVE Part of the German report on Churchill Mark I, II and III tanks captured at Dieppe. (Author's collection)

in flames. After the battle this tank was found to have been hit 38 times on the front. One of these rounds had penetrated the armour. There were impacts from high-explosive rounds and from 75mm armour-piercing rounds, one of which had penetrated. There were also eight impacts on the rear of the tank from 57mm rounds – the calibre of the British and Allied 6-pounder anti-tank guns used by the infantry. Given that the hits were in the rear of the tank,

BELOW Churchill Mark III with Kingforce showing 6-pounder shot being loaded. Note the primer protecting clip on the base of the round on the turret roof and shot damage to front of turret, side of turret and to track guards. The man at the front is using wire cutters to remove wire caught up in the tracks. (Tank Museum)

it is possible that they came from British or Australian manned guns as the tank reversed towards them. The tank burned out. Lieutenant Appleby was killed and is commemorated on the Alamein Memorial in Egypt (on Column 13).

Seven days later the five remaining tanks of Kingforce returned to battle. On one tank the traverse mechanism for its turret failed. Two more tanks were hit (one by no fewer than nine 50mm rounds), which prevented their turrets from traversing. Another was hit 30 times and lost a track. Major King's tank had a very lucky escape: a 50mm round came through the open driver's vision port but caused no injuries and did not disable the tank. Although the five tanks were recovered and returned to working order, this marked the last action of Kingforce as a unit.

The war in Tunisia

The next, and much more substantial, Churchill tank actions took place early in 1943 not far away, in Tunisia, where an Allied army under General Eisenhower had landed to fight the German and Italian armies, which were dug into well-established defensive positions among hills and mountains.

The 25th Army Tank Brigade was chosen to join these battles, comprising three Churchill-equipped regiments: North Irish Horse, 51 Royal Tank Regiment (RTR) and 142 Regiment Royal Armoured Corps. Shortly afterwards the 21st Army Tank Brigade joined them. This too was equipped with Churchill tanks, and comprised 12 RTR, 48 RTR and 145 Regiment RAC (formed from the 8th Battalion the Duke of Wellington's Regiment).

It was in Tunisia that the Churchills first met, and defeated, the Tiger tank and the 88mm anti-tank gun. The tanks and their crews acquitted themselves superbly and showed the value of this tough infantry tank as a complement to the infantrymen on the ground, as these contemporary comments show:

'So far, reports tend to show that the morale effect on the enemy is considerable. The Churchill surprised the Germans. As the surprise wears off, so, doubtless will the effect to a degree.

'The following is an extract from an official interrogation of a P.W. [prisoner of war], taken during the engagements at El Aroussa on 27–28 February 1943. Asked to account for the fiasco of February 26, P.W. said 'It was the tanks. We knew of course that you had some tanks here, what we did not know was that they were Churchills. That's what upset our calculations.'

Lieutenant Peter Gudgin gave a wonderfully laconic description of battle damage from the same campaign:

'Churchill IV No T68605 commanded by myself, received 4 hits from German A/Tk guns of various calibres. Hits 2, 3 and 4 occurred practically simultaneously, about ten seconds after hit 1. The crew commenced to bale out immediately after the first hit, but were not clear before the last three.

'Hit 1. An 88m/m shot hit the co-driver's M.G. mounting, penetrated and passed down the length of the tank into the engine. En route, it set off the four rounds of 6 pdr ammunition in unarmoured clips behind the driver and started a fire in the engine compartment. The fire in the engine compartment was extinguished by the CO_2 bottle but later re-ignited by ammunition fire. The tank stopped, the driver and co-driver were the only members of the crew affected by this round, and commenced to bale out.

'Hit 2. A 75m/m shot hit the left hand side of the turret at approximately 45 degrees to normal, beside the Commander, but did not penetrate. Petals of armour, however, set off a 6 pdr round which had been placed in the grenade rack (the grenades having been removed). Of this I am certain because I was beside the round when it went off.

'Hit 3. One 50m/m shot hit the right hand front of the turret ring, stuck in the armour with its nose just through, and set off the 2" smoke ammunition. I know this happened immediately after impact because I smelt the different smoke.

'Hit 4. Another 50 m/m shot hit the right hand side plate behind the escape hatch at approximately 45 degrees to normal, did not completely penetrate, but flaking inside set off several rounds near the operator.

'Casualties – Driver: burned.
Co-driver: one arm and three ribs broken and burned.
Operator: splinters in leg, one arm broken by hit on turret.
Gunner: slightly burned on back, splinters in leg.
Commander: blinded by flash, splinters in leg.

'I had the opportunity of making a detailed examination of the tank in Bde. Workshops where I was able to obtain the above details, as at the time smoke filled the tank and prevented observation.

'The ammunition carried in the bin behind the turret [the four rounds behind the operator which sat on the hull floor rather than the turret floor] was intact, but all other ammunition had burnt. The tank carried no 6 pdr H.E.'

A few stories from the battles may give some sense of this phase of the war.

In an action on 5 February 1943, at a place known as Steamroller Farm, 13 Churchills from A Squadron 51 RTR fought along a pass beside two companies of the 2nd Battalion Coldstream Guards. Against them were two battalions of Hermann Goering Regiment as well as elements from a Panzer Grenadier Regiment – very tough and capable opponents. With the infantry pinned down, orders were given to clear the pass at any cost. Two Churchills commanded by Captain E.D. Hollands and Lieutenant J.G. Renton made a 1,500yd advance across exposed ground covered by an 88mm gun, which they charged down. The two Churchills then climbed the steep hill at the end of the pass to engage a more distant enemy force that certainly never expected to be fired on from such a height. They destroyed two more 88mm guns, two 50mm guns, 25 wheeled vehicles, two mortars and two Panzer III tanks.

On 27 February 1943, Lieutenant Hern of the North Irish Horse fired at a tank at 500yd range and found that he had knocked out a Tiger.

Another action, involving 48 RTR, illustrates the confusion of battle and the toughness

of the Churchill. In April 1943, B Squadron was supporting a battalion of the East Surrey Regiment against German tanks on a ridge. These tanks were engaged at some 200yd range. One Churchill was that commanded by Lieutenant Peter Gudgin, whose report is mentioned earlier. In this frantic close-range encounter the Churchills accounted for three Panzer IIIs, one Panzer IV and one Tiger tank. It seems that these German tanks, apart from one that had brewed up, were abandoned

ABOVE After taking Toukabeur front, British infantry assisted by Churchill tanks push on towards 'Longstop Hill' and the surrounding heights. This part of the country is extremely difficult for vehicles but the Churchill tank showed remarkable dependability in climbing the steep gradients, some of which were in the region of 1 in 3. *(Tank Museum)*

LEFT The Tiger tank knocked out by 48 RTR in April 1943. The incoming 6-pounder round bounced off the bottom of the barrel before glancing down from the mantlet. The turret was facing forwards when hit but was traversed to make the damage easier to photograph. *(Author's collection)*

by their crews, in some cases without being badly impaired. The Tiger was damaged by a 6-pounder round that quite probably stopped its turret from traversing, and may have injured one or more of the crew by deflecting downwards off the mantlet on to the thin armour above the driver and front gunner. This Tiger was returned to Britain for investigation – by a team that ironically included Peter Gudgin – and can now be seen, still with its battle damage, at the Tank Museum in Bovington.

Churchill tanks fought well in Tunisia, vindicating those who had advocated their design and use in the support of infantry. The tank's ability to climb steep slopes led to it being affectionately called the 'Mountain Goat', and was an important element in its success.

BELOW Personal message from the Army Commander, 'Monty', to mark the end of the war in Tunisia. (Author's collection)

The war in Italy

The Churchill tanks went from North Africa to the war in Italy, with the 25th Army Tank Brigade leaving North Africa in April 1944 and the 21st Army Tank Brigade the following month. Their battles, not least for the Hitler Line and the Gothic Line, are well documented, as are the instances of truly astonishing bravery in these attacks on long-prepared German defensive positions.

The battle for the Hitler Line illustrates the toughness of the Churchill and its legendary climbing ability. The battle took place in May 1944 and included the North Irish Horse in support of the Canadians of Princess Patricia's Infantry. The German defences were very well organised, and had been prepared long in advance. They included at least eight Panther tank turrets dismounted and set on concrete bases and accompanied by machine guns, rocket launchers and self-propelled guns. Bunkers and other defensive positions were carefully sited and trees cut to give fields of fire on terrain that was prepared with killing zones into which attacking tanks had to move. Snipers were deployed, many strapped to trees and, according to accounts of the battle, more than willing to fire on stretcher-bearers as well as combatants.

In the minutes before the tanks and infantry left their start line, some 786 Allied guns opened fire, with a creeping barrage across 3,200yd and to a depth of 3,000yd. This barrage was meant to move ahead of the advancing tanks and infantry – the timing was intended to be five minutes for the first 100yd of the attack, and later three minutes per 100yd.

However, the Germans responded by shelling the ground behind the Allied barrage and the effect was not only one of lethal fragments from air bursts and shells exploding on the ground, but also intense and deafening noise. All this kicked up so much dust that there was no visibility for tank commanders and drivers, who could not readily open up their hatches and direct their tanks in such intense firing. With only their periscopes to see through, and dust everywhere, control of the tanks in battle became almost impossible. As often happened, each tank was very much on

LEFT Infantry carry
mortar ammunition
over the river Senio
floodbank. A Churchill
Crocodile can be seen
through the smoke
in the background.
*(Jonathan Falconer
collection)*

its own, and each crew depended on its wits
to survive. Worse still, damage to tank aerials
and terrible casualties among the infantry
radio operators meant that all communication
between tanks and infantry failed.

In A Squadron, North Irish Horse, Lieutenant
Donald Hunt was in command of 5 Troop, on
the right flank of the whole battle, and they
went forward at 6:00am into the German
counter-barrage, with some 1,000yd between
the start line and their objective. It may give
some idea of the conditions to realise that
after an hour and a half they had only moved
some 500yd. The squadron was held up by
a minefield and the squadron leader, Major
Griffith, using his tank radio, asked Lieutenant
Hunt to seek a way around the minefield on
the right of the advance. Lieutenant Hunt took
his tank to the right of the troop, and indeed
of the whole battlefield, to probe forward.
The Germans had designed the killing zones
well and, unable to see what lay beside him,
Lieutenant Hunt's tank fell into an enormous,
almost conical, valley, several hundred feet
deep, turning over and landing first on its roof

and then rolling on to its tracks on a ledge
some way down the side.

Amazingly the crew survived, as did the tank.
As the fall had knocked out the wireless sets,
Lieutenant Hunt had to make his way back on
foot through the battle to report to Major Griffith,
both on the impassable obstacle and the fate of
his tank. The battle was at a critical phase and
every tank was needed; Major Griffith ordered
him to try to get his tank out of the valley and
back into the fight.

Lieutenant Hunt ran back under sniper fire
and, surveying the valley, decided that if the
tank would run, it might just be able to reach
the bottom and climb the steep side opposite.
This it did, under the superb driving of Davy
Graham, making the final ascent after several
failed attempts and with the crew certain that it
would topple over backwards.

That the tank and crew survived a rolling fall
of some 40ft on to a ledge, and that they were
able and willing to risk all to drive out again and
re-engage in the battle, is remarkable and a real
testament to the toughness of both men and
machine.

RIGHT The cap badge of the North Irish Horse. *(Author's collection)*

LEFT This is the epaulette of a soldier of 25th Army Tank Brigade, showing the maple leaf emblem awarded after the Hitler Line battle. *(Author's collection)*

BELOW This Special Order of the Day marked the end of the fighting in Italy on 2 May 1945. *(Author's collection)*

ALLIED FORCE HEADQUARTERS
2 May, 1945

SPECIAL ORDER OF THE DAY

Soldiers, Sailors and Airmen of the Allied Forces in the Mediterranean Theatre

After nearly two years of hard and continuous fighting which started in Sicily in the summer of 1943, you stand today as the victors of the Italian Campaign.

You have won a victory which has ended in the complete and utter rout of the German armed forces in the Mediterranean. By clearing Italy of the last Nazi aggressor, you have liberated a country of over 40,000,000 people.

Today the remnants of a once proud Army have laid down their arms to you—close on a million men with all their arms, equipment and impedimenta.

You may well be proud of this great and victorious campaign which will long live in history as one of the greatest and most successful ever waged.

No praise is high enough for you sailors, soldiers, airmen and workers of the United Forces in Italy for your magnificent triumph.

My gratitude to you and my admiration is unbounded and only equalled by the pride which is mine in being your Commander-in-Chief.

H.R. Alexander.

Field-Marshal,
Supreme Allied Commander,
Mediterranean Theatre.

The North Irish Horse can trace its origins back to 1901 and saw action through the First World War, having been (along with the South Irish Horse), the first non-regular British troops to arrive in France and see action. Following the Armistice, the North Irish Horse went into suspension until 1939 when it became part of the Royal Armoured Corps and, in 1941, an infantry tank unit equipped with the Churchill tanks.

In 1943, as part of the 25th Army Tank Brigade, the regiment sailed for Algeria. It went straight into action at Hunt's Gap and Sedjenane and fought in all the ensuing battles, ending that campaign by being among the first tanks to enter Tunis on 8 May 1943.

In April 1944 the regiment joined the Eighth Army in Italy. Its first major action was in support of Canadian infantry in the attack on the Hitler Line.

In the Hitler Line battle 34 officers and men were killed, with 36 wounded; 25 tanks were also put out of action. The Canadians lost over 1,000 killed, wounded and missing. In recognition of the support given to the Canadians in the battle, General G.C. Vokes, DSO, commanding 1 Canadian Infantry Division, passed on the message 'that he would be pleased if all ranks 25 Tank Brigade would wear a Maple Leaf emblem in token of the part played by the Brigade in assisting 1 Canadian Infantry Division to breach the Adolph Hitler Line'. The North Irish Horse, as part of 25 Tank Brigade, adopted the maple leaf as a battle honour to be worn on the sleeve of their battledress. The maple leaf was also shown on the North Irish tank markings from this point.

Following the breaking of the Hitler Line, the regiment fought northwards until the German surrender in Italy.

The North Irish Horse still serves today, as a yeomanry squadron, in Northern Ireland, and proudly has two Churchill tanks on display.

Message from the Commander-in-Chief to mark the start of the Normandy invasion. *(Author's collection)*

The campaign in north-west Europe

From Italy the story of the Churchill tank's involvement in the Second World War must move back in time to the Normandy landings and the war in north-west Europe. By June 1944 the Churchill was a proven tank with an outstanding record; it was also very adaptable and provided the basis for a significant number of special-purpose vehicles in support of the invasion, which are described in Chapter 2.

The lessons of Dieppe, and the fruits of much planning and brave reconnaissance, enabled the British landings to deploy these specialist Churchills right from the start of the campaign, and they proved their worth all the way into Germany and to the end of the war.

By the time of the Normandy landings the early Churchills, the Marks I and II, had been upgraded to later specifications with new turrets. The main types in service were the Marks III and IV, with the Mark VII in gun tank and Crocodile form making welcome appearances. Many of the Marks III and IV tanks were 'uparmoured' by the addition of appliqué armour on the sides of

21 ARMY GROUP

PERSONAL MESSAGE FROM THE C-in-C

To be read out to all Troops

1. The time has come to deal the enemy a terrific blow in Western Europe.

   The blow will be struck by the combined sea, land, and air forces of the Allies—together constituting one great Allied team, under the supreme command of General Eisenhower.

2. On the eve of this great adventure I send my best wishes to every soldier in the Allied team.

   To us is given the honour of striking a blow for freedom which will live in history; and in the better days that lie ahead men will speak with pride of our doings. We have a great and a righteous cause.

   Let us pray that "The Lord Mighty in Battle" will go forth with our armies, and that His special providence will aid us in the struggle.

3. I want every soldier to know that I have complete confidence in the successful outcome of the operations that we are now about to begin.

   With stout hearts, and with enthusiasm for the contest, let us go forward to victory.

4. And, as we enter the battle, let us recall the words of a famous soldier spoken many years ago :—

   "He either fears his fate too much,
   Or his deserts are small,
   Who dare not put it to the touch,
   To win or lose it all."

5. Good luck to each one of you. And good hunting on the mainland of Europe.

   B. L. Montgomery
   General
   C-in-C 21 Army Group.

1944.

BELOW **This is the side of a Mark IV tank, now a gate guardian with the North Irish Horse in Belfast. It shows the appliqué armour added to the sides (and in the case of the Mark III, to the turret front). The circular cut-outs make space for the large conical nuts holding the hull sides to the inner plate. The pannier door has a 'T' shaped piece of appliqué armour fitted. This tank has an interesting history in that it was in service with the Irish Army after the war, but when no longer needed it was buried. In more recent years it was dug up and kindly donated by the Irish Government and Defence Forces to the North Irish Horse.** *(Author's collection)*

BELOW **Much time was spent practising beach landings and learning from the lessons of Dieppe.** *(Tank Museum)*

RIGHT The instruction pamphlet for the *Panzerfaust*, showing in this case a Russian tank rather than a Churchill. *(Author's collection)*

BELOW A Churchill Mark IV or VI with 75mm gun from 7 RTR, 31st Tank Brigade, supporting infantry of 8th Royal Scots during Operation 'Epsom', 28 June 1944. Note the extra stowage hung on the turret and the cleaning stores for the main gun on the back box. *(Tank Museum)*

the hull and the sides and front of the Mark III turret. In this form, with a 75mm gun, the Mark III was termed the Mark III*.

As well as taking part in advances, with all the hazards of unseen enemy tank and anti-tank guns, mortars, snipers and *Panzerfausts* (the German infantry's equivalent of the bazooka), the Churchills often had to hold ground under constant fire, notably in the hard-fought battles on Hill 112 in Normandy. One Churchill regiment, 107 Regiment (The King's Own), part of 34 Tank Brigade, were positioned on Hill 112 for ten days in the battle, just below the crest.

The discomfort of the crews in these situations is hard to imagine, and no photographs of the inside of the tanks can convey the hardship and stress affecting the crews. Such was the constant bombardment from the Germans that the crews hardly dared to move outside their tanks for more than a few minutes. This meant that they had to do almost everything inside the tanks – sleeping, cooking, eating, washing and shaving. Some crews chose to sleep under their machines, but many would not, as tank crews had always heard, and feared, stories of tanks sinking in soft ground and crushing those underneath. When nature's call became unavoidable, the crew would dash to a nearby shell crater, tense and ready at any moment to rush back to their tanks if mortars or shells came in. For lesser needs, empty shell cases were used for urinating in, and the contents dumped through a hatch.

The effect of standing in the tanks for prolonged periods also led to swollen feet – so much so that one officer reported having to wear plimsolls for some days afterwards as he could not fit into his boots.

Fittingly, there is now a memorial on Hill 112 with a Churchill Mark VII tank to commemorate the fighting there.

Another famous regiment with Churchill tanks in Normandy was the 9th Battalion Royal Tank Regiment, which fought from soon after D-Day until the end of the war in Europe. They made their landings after crossing in the Channel storms of 18–20 June 1944 and fought across Holland into Germany, also assisting in holding the German counter-attack in the Ardennes.

Bill Thompson in 9 RTR was the loader/wireless operator in a Churchill named *Impassive*, commanded by Corporal Freddie Horner. This tank was part of 11 Troop, which was, in turn, commanded by Eb Wood. This is Bill Thompson's account of an encounter that took place on 29 October 1944 near Roosendaal in Holland:

We were fighting our way over bogs and dykes in pouring rain and came to open country. Our tank moved alongside a farm outbuilding with the other two tanks of our troop to the rear. We had been there only a matter of minutes when through the periscope I saw a flash about a mile away and was sure it was a German '88. There was a second shot which hit the ground next to the tank; it became obvious that it was an '88 and he was finding the range.

Without hesitation our tank commander Freddie Horner instructed our driver to

reverse out as quickly as possible, which he did. Over the B Set [part of the Wireless No 19 – see section on 'Radios and communications in the Churchill' in Chapter 4] came our troop leader Mr Wood telling us to get back into position. When we didn't move, he moved his tank into exactly

9TH BATTALION ROYAL TANK REGIMENT

The 9th Battalion Royal Tank Regiment (9 RTR) had its origins in the formation, in December 1916, of the 9th Battalion Heavy Machine Gun Corps, which served at the Battle of Cambrai. The battalion fought to good effect through the rest of the First World War and was awarded the Croix de Guerre avec Palme, and also the honour of wearing the badge of the French 3rd Division commanded by General Bourgon. This badge was worn on the sleeve and was retained by 9 RTR in the Second World War when they also took its words as their

motto: *'Qui s'y frotte, s'y brule'*, meaning 'Who touches me burns'. The battalion had been disbanded after the First World War and re-formed in 1941, to become one of the first armoured regiments to receive the Churchill tank, in which they served with distinction to the end of the war.

Their wartime story is told in Peter Beale's excellent book, *Tank Tracks, 9th Battalion Royal Tank Regiment at War 1940–45*, from which the story of their encounter with an 88mm gun in this chapter is taken, with Peter Beale's kind permission.

RIGHT From an honour dating back to the First World War, the men of 9 RTR wore a replica of the badge of General Bourgon's French 3rd Division on their sleeves. The motto, *'Qui s'y frotte, s'y brule'*, means 'Who touches me burns'. *(Author's collection)*

ABOVE The cap badge of the Royal Tank Regiment. *(Author's collection)*

LEFT This humorous Christmas card from 9 RTR shows the progress of their tanks from D-Day. *(Author's collection)*

the position we had been alongside the outbuilding. Within seconds there was one almighty bang and he had taken a direct hit on the front of his turret. When the dust had settled and he had made a dazed but hurried retreat it was discovered that an AP shot had penetrated [but not actually gone all the way through the armour] into the turret, just as though it had been drilled by a huge drill.

Mr Wood's tank [was] a new Mark VII. Ours was a Mark IV [which, of course, had much thinner frontal armour]. Had we not moved quickly from our position the AP shot would surely have holed our turret and killed the turret crew, and possibly blown the tank up completely had the ammunition been struck. Needless to say we were forgiven for moving from our original position. Mr Wood's tank

BELOW A personal message from 'Monty' to commemorate the end of war in Europe. *(Author's collection)*

went back to the REME workshop for repair and was soon returned ready for action again.

In fact the Churchill could take a lot of punishment. The 34 Armoured Brigade History recorded: 'One notable tank casualty was a Mark VII of 9 RTR which sustained 9 direct hits in front from 75mm AP shot at short range without being completely penetrated by any!'

After the Second World War, the story of the Churchill tank in British Army service moves on five years to Korea, for what was to be its last use in active war service.

The Korean War

On the morning of 25 June 1950 the North Koreans attacked South Korea across the 38th Parallel. Following this, the United Nations adopted a resolution calling for the North Koreans to withdraw. They did not do so, and on 30 June President Truman ordered US forces to come to the aid of South Korea and the UN Security Council called on members to provide support.

On 29 July, the 7th Royal Tank Regiment – which had served in Churchill tanks during the war – was ordered to mobilise a squadron to serve in Korea in Crocodile flame-thrower tanks. C Squadron, at the time in Norfolk guarding an airfield, was chosen for the role. The squadron was to be commanded by Major A.J.D. Pettingell. Half of their tanks were in Germany where they had been standing in vehicle parks for some five years. By 12 October the squadron had mustered men and tanks, carried out some training and was on a ship to the Far East.

This was to be a difficult role, for neither the men nor the machines were as prepared as might have been wished, but the Churchill was still able to carry out its role of supporting the infantry.

There was relatively little fighting for the Churchills, but the squadron had to maintain its readiness in appallingly cold conditions, with few spares and little backup. Two of the tanks (one Crocodile and one ARV) had to be destroyed to prevent capture. The need to do this was learned the hard way after the squadron had come under fire from a captured Cromwell tank and had to knock this out with help from some Centurions, so there was every

21 ARMY GROUP

PERSONAL MESSAGE FROM THE C-IN-C

(To be read out to all Troops)

1. On this day of victory in Europe I feel I would like to speak to all who have served and fought with me during the last few years. What I have to say is very simple, and quite short.

2. I would ask you all to remember those of our comrades who fell in the struggle. They gave their lives that others might have freedom, and no man can do more than that. I believe that He would say to each one of them:

   "Well done, thou good and faithful servant."

3. And we who remain have seen the thing through to the end; we all have a feeling of great joy and thankfulness that we have been preserved to see this day.

   We must remember to give the praise and thankfulness where it is due:

   "This is the Lord's doing, and it is marvellous in our eyes."

4. In the early days of this war the British Empire stood alone against the combined might of the axis powers. And during those days we suffered some great disasters; but we stood firm: on the defensive, but striking blows where we could. Later we were joined by Russia and America; and from then onwards the end was in no doubt. Let us never forget what we owe to our Russian and American allies; this great allied team has achieved much in war; may it achieve even more in peace.

5. Without doubt, great problems lie ahead; the world will not recover quickly from the upheaval that has taken place; there is much work for each one of us.

   I would say that we must face up to that work with the same fortitude that we faced up to the worst days of this war. It may be that some difficult times lie ahead for our country, and for each one of us personally. If it happens thus, then our discipline will pull us through; but we must remember that the best discipline implies the subordination of self for the benefit of the community.

6. It has been a privilege and an honour to command this great British Empire team in western Europe. Few commanders can have had such loyal service as you have given me. I thank each one of you from the bottom of my heart.

7. And so let us embark on what lies ahead full of joy and optimism. We have won the German war. Let us now win the peace.

8. Good luck to you all, wherever you may be.

B. L. Montgomery

Germany,
May, 1945.

Field-Marshal,
C.-in-C.,
21 Army Group.

reason to prevent the vehicles' capture if they broke down and could not be repaired.

An extract from a report to the officer commanding 7 RTR gives a flavour of daily life:

The [tanks] have done between 350 and 500 miles depending on Troops. We have had two pistons and conrods through the sump in the last two days. Have hastily changed [engine oil].

Churchills require an easily opened drainage valve in the engine and gearbox compartment so that water picked up in fording can be easily got rid of. Two inches of water in the bottom can freeze around gear selector rods and seize them solid until the engine heat melts the ice – up to 30 minutes. If the tank has been left overnight in gear, this means holding the clutch pedal down for 30 minutes – excessive clutch wear results. We chock up the bogies and leave the tanks in neutral whenever possible.

Turret rings seize pretty hard due to frozen grease. Solution – lubricate with OE 30 and place straw around the turret ring outside.

Crocodiles on icy roads are one thing that terrifies everyone in Korea. One skidded sideways up to 45 feet on the recent advance to Yongdeungpo.

ARVs – a sad tale! Due to the tracks skidding, they can't pull the skin off a rice pudding as far as towing is concerned. The winch and spade combination, however, is still supreme among allied recovery kit. But both of ours broke the drive to the scavenge pump and the petrol pumps within four miles of each other. One was recovered and one had to be burnt.

Demolition. A hard and fast SOP [Standard Operating Procedure] for brewing abandoned tanks is essential. H.E. rounds in the breech and muzzle and 45 yards of signal cable deals with the gun. [The writer means firing the gun by pulling on the cable attached to the firing mechanism (from a safe distance of 45yd), so that the round in the breech meets the round stuck in the muzzle and blows up the barrel.] But a certain method of brewing up when there is only a little petrol in the tank and only a few rounds of ammo left presents problems.

Sgt. Robbins' tank, which we were unable to recover when the panic button was pressed at Seoul [the urgent evacuation of the city] was dealt with by 15lbs of US Satchel Charge in the driver's compartment and 30lbs in the turret. Result – one entire sideplate across the road embedded in a wall, turret off and hull split in two, and one the neatest cross sections of a Churchill engine yet seen in Korea, one bogie assembly blown across the square 147 paces away.

With the stalemate that followed, the Churchill tanks' last active service role in the British Army ended, although the specials continued to be seen on tank parks for many years into the 1960s. Some of these are described in Chapter 2.

All in all, it was a long and successful role for a tank of which so little was expected at the outset, and the Churchill rightly remains an icon of its time.

ABOVE A Churchill Mark VII Crocodile of 7 RTR shells Seoul in February 1951 from across the Yongdeungpo. The tank bears the unusual name of *Gynaecolater*. *(Author's collection)*

BELOW 'Still supreme among allied recovery kit', a Churchill ARV salvages an armoured car in Korea, which appears to have driven over a mine and lost its rear wheel station. *(Tank Museum)*

MARK IV CHURCHILL.   (A.22.C.   CAST TURRET.

R. GUN).

Chapter Four

Anatomy of the Churchill

The Churchill may have looked like a First World War tank, but it was spacious inside and reliable once established in service. A great deal of evolution was packed into a very short period, but the tank remained recognisable throughout and its essential features endured.

OPPOSITE A wonderfully detailed drawing of the inside of a Churchill Mark IV tank with a long-barrelled 6-pounder that has free elevation. *(Author's collection)*.

77

ANATOMY OF THE CHURCHILL

This chapter concentrates on the gun tanks – the backbone of the Churchill-equipped regiments. Even these tanks were modified in the field and, as can be seen in contemporary photographs, varied in their set-up from theatre to theatre. Below are descriptions of the main variants. Interestingly, and somewhat confusingly, basic statistics for the tanks and their major components vary in different official publications. Where facts and figures are set out below, they are drawn from official data and tables.

TERMS EXPLAINED

For those who are not familiar with tanks, and Churchills in particular, these diagrams, and a few words of explanation may make the book easier to follow. As explained in Chapter 1, the tanks evolved in a series of 'Marks', which are referred to by Roman numerals starting with the earliest, I, and ending with VII and VIII. For brevity's sake, Churchills are referred to by mark in much of the text. The armament they used also progressed, from a relatively tiny 2in (40mm) gun to other, larger and more effective, weapons. The guns are described with a mixture of terminology. Some are referred to by the weight of the projectile: the 2-pounder and the 6-pounder; the 'pounder' is often abbreviated to 'pr' or 'pdr'. The rest are described by reference to the calibre, that is to say the diameter, of the projectile: 75mm, 95mm and 3in. At the time, the use of a mixture of imperial and metric measurements was common, if a little confusing.

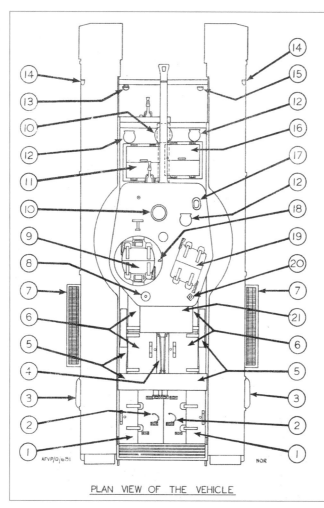

PLAN VIEW OF THE VEHICLE

1. Gearbox compartment hatches.
2. Hatch safety chains.
3. Spare track link (sometimes referred to as a 'track plate'.
4. Hatch support rods to prevent the hatches from falling when open.
5. Exhaust pipes and silencers.
6. Engine compartment hatches.
7. Air inlet louvres.
8. 'A' set aerial mounting.
9. Commander's cupola.
10. Covers over ventilation fans.
11. Front gunner's hatches.
12. Periscopes.
13. Masked headlamp, sometimes called a 'blackout' lamp.
14. Side lamps, mounted on the track guards. The projecting track guards are often referred to as the 'horns' where they stick out in front of the hull.
15. Ultraviolet headlamp.
16. Driver's hatches.
17. 2in bomb-thrower aperture.
18. Spotlight mounting bracket.
19. Loader/operator's hatches.
20. 'B' set aerial mounting.
21. Turret stowage bin.

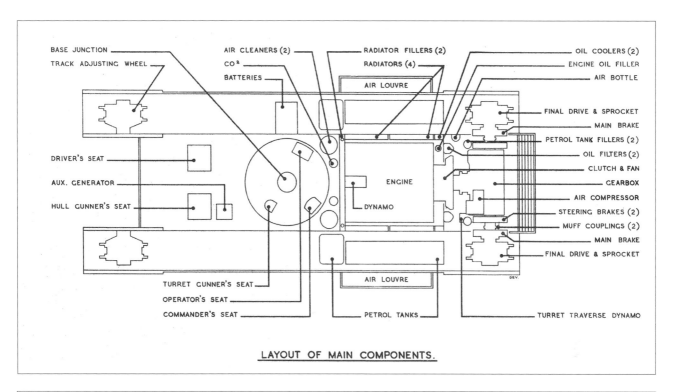

LAYOUT OF MAIN COMPONENTS.

Labels (clockwise from top left):
BASE JUNCTION · TRACK ADJUSTING WHEEL · AIR CLEANERS (2) · CO² · BATTERIES · RADIATOR FILLERS (2) · RADIATORS (4) · AIR LOUVRE · OIL COOLERS (2) · ENGINE OIL FILLER · AIR BOTTLE · FINAL DRIVE & SPROCKET · MAIN BRAKE · PETROL TANK FILLERS (2) · OIL FILTERS (2) · CLUTCH & FAN · GEARBOX · AIR COMPRESSOR · STEERING BRAKES (2) · MUFF COUPLINGS (2) · MAIN BRAKE · FINAL DRIVE & SPROCKET · TURRET TRAVERSE DYNAMO · PETROL TANKS · COMMANDER'S SEAT · OPERATOR'S SEAT · TURRET GUNNER'S SEAT · HULL GUNNER'S SEAT · AUX. GENERATOR · DRIVER'S SEAT · ENGINE · DYNAMO

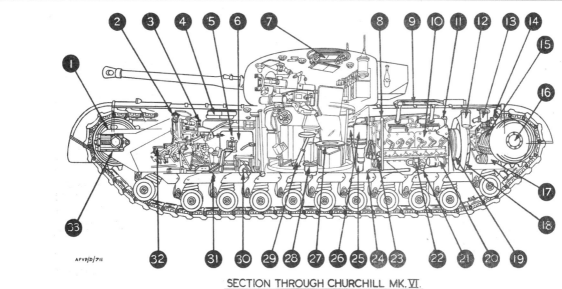

AFVD/D/7II

SECTION THROUGH CHURCHILL MK. VI.

1. Front idler.
2. Front gunner's periscope.
3. Driver's periscope.
4. Driver's and front gunner's hatches.
5. Pistol port (sometimes referred to as a 'revolver' port).
6. Pannier door.
7. Commander's cupola.
8. Radiator filler pipe.
9. Exhaust pipes.
10. Engine.
11. Engine oil filler.
12. Engine oil filler.
13. Petrol tank filler.
14. Air compressor.
15. Gearbox.
16. Final drive.
17. Turret traverse dynamo.
18. Fan, sometimes known as the 'Sirocco' fan.
19. Clutch.
20. Starter motor.
21. Petrol pump.
22. Engine oil scavenge pump.
23. Water pump.
24. Radiator drain tap.
25. CO2 bottle.
26. Engine air cleaners.
27. Commander's pedestal.
28. Rotary base junction.
29. Turret gunner's seat.
30. Auxiliary generator (Tiny Tim or Delco Remy).
31. Hull gunner's seat.
32. Hull gunner's steering bar.
33. Track adjusting assembly.

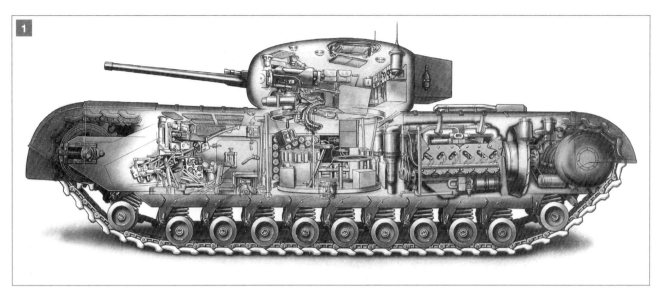

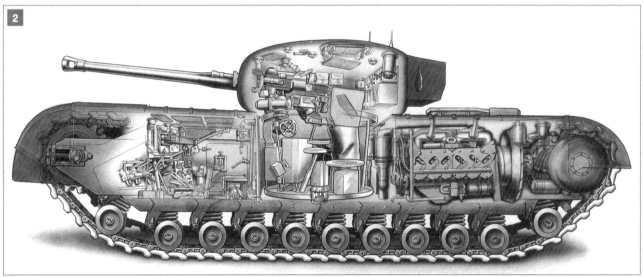

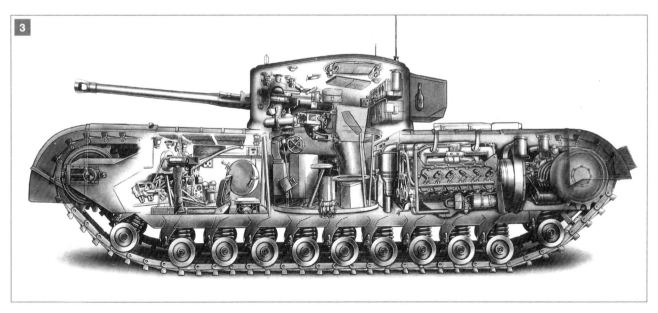

1 A Churchill Mark IV. The division into driving, fighting, engine and gearbox compartments can clearly be seen. The gun is a 6-pounder with free elevation (note the 'C' shaped bracket for the gunner's shoulder). *(Tank Museum)*

2 A Mark VI in cross-section. This tank has a 75mm gun with geared elevation (note the hand-wheel under the gun), but an early two-periscope cupola. If you compare the ammunition stowage with the diagram of the Mark IV, you will see that this Mark VI has a tall armoured stowage bin on the turret floor, the other side of the gunner's seat. Although not shown here, it would presumably have had armoured 'cupboard' stowage for the remaining 75mm ammunition. Another difference is that the coaxial Besa has a metal chute (as did the Mark VII) for spent cartridges and belts. *(Author's collection)*

3 A Mark VII in cross-section. Note the round pannier door. This appears to be an early Mark VII as it does not have the all-round vision cupola. *(Author's collection)*

Layout of the tank and major components

The gun tanks can all be regarded as having four compartments:

■ Front, housing the driver and front gunner.
■ Fighting, housing the turret crew.
■ Engine, behind the fighting compartment, separated from it by a steel bulkhead with opening partitions, housing the engine, petrol tanks, radiators and oil coolers.
■ Gearbox, housing the gearbox and compressor, steering and main brakes, with the dynamo for the turret traverse on the floor of the compartment.

Driver/front gunner compartment

This part of the tank is relatively spacious. The driver and front gunner sit upright with space to the sides and behind them. The seats are adjustable forwards and backwards and the early front gunner's seat has a cut-out behind his right shoulder. This is thought to originally have been

BELOW The driving compartment of a Mark VII is similar to the earlier tanks, but note the round vision port, two periscopes for the driver and the addition of a second handle for guiding the Besa, which actually has a remote trigger and hand grip on the end of it. Note also that the ignition stop switch has been moved to the centre of the roof. *(Author's collection)*

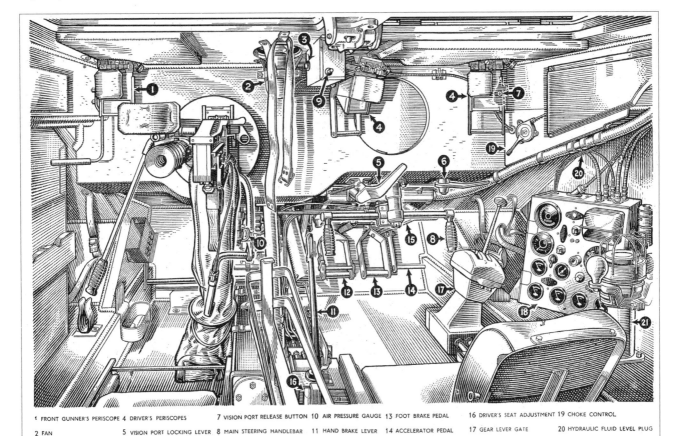

1 FRONT GUNNER'S PERISCOPE	4 DRIVER'S PERISCOPES	7 VISION PORT RELEASE BUTTON	10 AIR PRESSURE GAUGE	13 FOOT BRAKE PEDAL	16 DRIVER'S SEAT ADJUSTMENT	19 CHOKE CONTROL
2 FAN	5 VISION PORT LOCKING LEVER	8 MAIN STEERING HANDLEBAR	11 HAND BRAKE LEVER	14 ACCELERATOR PEDAL	17 GEAR LEVER GATE	20 HYDRAULIC FLUID LEVEL PLUG
3 FAN SWITCH	6 VISION PORT OPENING LEVER	9 IGNITION STOP SWITCH	12 CLUTCH PEDAL	15 THROTTLE HAND LEVER	18 INSTRUMENT PANEL	21 BINNACLE COMPASS

Fig. 11.   General view of controls from driver's seat.

RIGHT This is the driving compartment of a Mark IV, but in fact it would have looked similar in all Churchills from the Mark II to the Mark VI. The driver has a rectangular vision port and only one periscope. The area looks quite crowded when it is fully stowed. *(Author's collection)*

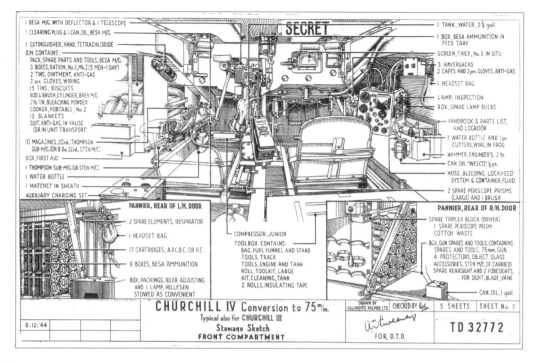

SECRET

1 BESA M/G WITH DEFLECTOR & 1 TELESCOPE
1 CLEARING PLUG & 1 CAN, OIL, BESA M/G.
1 EXTINGUISHER, HAND, TETRACHLORIDE
BIN CONTAINS:
PACK, SPARE PARTS AND TOOLS, BESA M/G.
3 BOXES, RATION, No.3, Mk.I (5 MEN-1 DAY)
2 TINS, OINTMENT, ANTI-GAS
2 prs. GLOVES, WIRING
15 TINS, BISCUITS
ROD & BRUSH, CYLINDER, BREN M/G.
2 lb. TIN, BLEACHING POWDER
COOKER, PORTABLE, No.2
15 BLANKETS
SUIT, ANTI-GAS, IN VALISE
(OR IN UNIT TRANSPORT)
10 MAGAZINES, 20 rd., THOMPSON
SUB-M/G.(OR 8 0a.32rd., STEN M/C.)
BOX, FIRST AID
1 THOMPSON SUB-M/G (OR STEN M/C)
1 WATER BOTTLE
1 MATCHET IN SHEATH
AUXILIARY CHARGING SET

1 TANK, WATER, 2¼ gall.
1 BOX, BESA AMMUNITION IN FEED TRAY
SCREEN, THICK, No. 5 IN SITU
3 HAVERSACKS
2 CAPES AND 2 prs. GLOVES, ANTI-GAS
1 HEADSET BAG
LAMP, INSPECTION
BOX, SPARE LAMP BULBS
HANDBOOK & PARTS LIST, AND LOGBOOK
1 WATER BOTTLE AND 1 pr. CUTTERS, WIRE, IN FROG
HAMMER, ENGINEER'S, 2 lb.
CAN, OIL, "WESCO", ½ pt.
HOSE, BLEEDING, LOCKHEED SYSTEM & CONTAINER, FLUID
2 SPARE PERISCOPE PRISMS (LARGE) AND 1 BRUSH

PANNIER, REAR OF L.H. DOOR
2 SPARE ELEMENTS, RESPIRATOR
1 HEADSET BAG
17 CARTRIDGES, A.P.C.B.C. OR H.E.
9 BOXES, BESA AMMUNITION
BOX, PACKINGS, IDLER ADJUSTING AND 1 LAMP, HELLESEN STOWED AS CONVENIENT

COMPRESSOR, JUNIOR
TOOLBOX CONTAINS:
BAG, FUEL FUNNEL AND STAND
TOOLS, TRACK
TOOLS, ENGINE AND TANK
ROLL, TOOLKIT, LARGE
KIT, CLEANING, TANK
2 ROLLS, INSULATING TAPE

PANNIER, REAR OF R.H. DOOR
SPARE TRIPLEX BLOCK (DRIVER)
1 SPARE PERISCOPE PRISM
COTTON WASTE
BOX, GUN SPARES AND TOOLS, CONTAINING:
SPARES AND TOOLS, 75mm. GUN
4 PROTECTORS, OBJECT GLASS
TOOLS, STEN M/C (IF CARRIED)
SPARE REARSIGHT AND 2 FORESIGHTS, FOR SIGHT, BLADE, VANE
CAN, OIL, 1 gall.

CHURCHILL IV Conversion to 75ᵐ/ₘ.
Typical also for CHURCHILL III
Stowage Sketch
FRONT COMPARTMENT

DRAWN BY ALLARDYCE PALMER LTD.   CHECKED BY   5 SHEETS   SHEET No. 5
6:12:'44   FOR D.T.D.   TD 32772

LEFT This is the only known contemporary photograph of the driver's compartment in a fully stowed Churchill. This is a Mark III or IV AVRE. The tall box in the middle of the photograph, nearest to the camera, holds the cartridges that set off the Petard Mortar. To its left is the box to catch spent cartridges from the coaxial Besa. The flared opening on top of this box normally held the bottom of the canvas chute from the Besa; it has been removed to allow the camera a better view. The front gunner did not have a proper seat; he had the cushion and a canvas strap that went around his waist, which can be seen draped over the Besa. *(Tank Museum)*

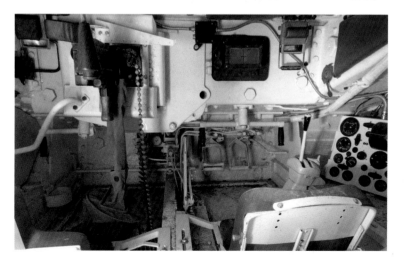

LEFT The driver's compartment of the Churchill Project's Mark IV. The green canvas chute under the Besa catches spent cartridges and belts. The Triplex vision block can be seen next to the driver's periscope. The gunner's sighting telescope is hidden by the drop lead for his headset. *(Author's collection)*

82

provided to allow room for the 3in howitzer. The front gunner's seat in pre-Mark VII (and some Mark VII) tanks had a small squab cushion at the top to enable him to sit 'head out' when safe to do so.

The driver and front gunner each have a pair of overhead hatches. These hatches can only be opened with the turret in certain positions. By the same token, the turret cannot traverse while the hatches are open. Beside and slightly behind the driver and front gunner are doors in the sides of the tank known as 'pannier doors'. The doors are very heavy and cannot easily be opened 'uphill' if the tank is parked on a slope with one side higher than the other. The edges of these doors were not painted, but instead kept in bare metal and lightly oiled to facilitate opening. On the pre-Mark VII tanks the pannier doors had 'pistol ports' – small hinged openings – although tried on prototype Mark VII tanks, these were dispensed with in production variants as being of no value and weakening the door.

In front of the driver is a 'vision port'. In the tanks before the Mark VII, this is rectangular and in two parts. The whole assembly is opened from the inside and can be partially closed leaving only the inner part open, so that the driver can see out through a 3in-thick Triplex block of armoured glass. On the Marks VII and VIII, the vision port was round and without an inner door and armoured glass block (although one prototype was produced with them). These vision ports, like the pannier doors, should not be painted around their edges. The driver in Marks I to VI had one periscope. In the Marks VII and VIII the driver had two periscopes, one to his right and the other to his left.

In early Churchill tanks the driver's and front gunner's periscopes could not see over the track guards to the sides, thus limiting their field of view. Mark VII and VIII tanks are found with an extended periscope and housing that enabled both the driver and front gunner to see over the track guards and these were retro-fitted to other, earlier, tanks.

Above the driver, in the corner of his overhead hatch, is the filler for the hydraulic oil

LEFT This is the inside of the Triplex armoured vision block. It has a cross etched on the front and back and is made up of layers of glass glued together. *(Author's collection)*

BELOW This is the back corner of the driver's hatch. A pipe covered with protective braid leads up to the filler for the hydraulic oil for the brakes, steering and clutch. To the right is the 'Distribution Box No 1' with the button for calling the commander. The driver's headphones and microphone plug into this. *(Author's collection)*

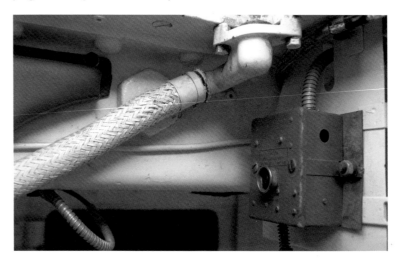

RIGHT The Besa
mount for the front
gunner in a Mark VII
and VIII. The gunner
holds the rod marked
'L' with his left hand
and with his right
hand uses 'R', the
mechanical trigger and
pistol grip on the right
side of the gun. 'A'
shows the ball mount
set in the thick frontal
armour of the tank.
(Author's collection)

BELOW Here is the
auxiliary generator
behind the front
gunner's seat. Note
the exhaust which
is pushed through
the pistol port in the
pannier door to vent
outside the tank.
(Author's collection)

that is used for the brake and steering. In front,
below the glacis plate, is the steering tiller and
the pedal cluster.

In front of the front gunner is a Besa machine
gun in an armoured gimbal mount, with a
telescope to the left of the gun. A canvas screen
beside the telescope safeguards the gunner's
face from debris and splash from the gun, as his
cheek is almost next to the ejecting cartridges.
He has a periscope to the left. The gimbal mount
enables the armoured housing for the Besa to

move left and right and up and down to a limited
extent. The telescope used was the same as
that for the main guns – including the graticule
markings for 6-pounder or 75mm. The front
gunner's periscope has a different handle and
larger, brow pad and handle from the others in
the tank. In all cases there is a lever fixed on the
left of the Besa mount, to help the front gunner
aim the Besa using his left hand, while his right
hand is on the pistol grip. In the later Marks VII
and VIII, there is a second lever on the right of
the gun with a remote mechanical trigger to fire
the Besa. Below the Besa is a canvas catcher,
clipped to the underside of the gun, for the spent
cases and belts that would otherwise rattle round
the hull. In all but late marks, the front gunner
has a single steering lever – an extension of the
driver's dual tiller. This enables him to steer the
tank if the driver is incapacitated. He also has a
'stop' button, which allows him to switch off the
ignition in an emergency.

Behind the front gunner's seat is a small
petrol generator. This was to charge the
batteries when the engine was not running.
The radios used a lot of power, and as the
tank's two 6V batteries (connected in series to
produce 12V) served the engine, the radios and
the lights, they could easily be run down when
the tank was stopped with the radios working.
The generator (either a 'Tiny Tim' or a 'Delco
Remy') had its own petrol supply piped from the
engine compartment under the turret floor, and

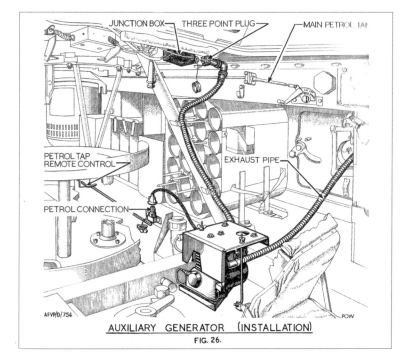

JUNCTION BOX — THREE POINT PLUG — — MAIN PETROL TAP

PETROL TAP
REMOTE CONTROL —

EXHAUST PIPE —

PETROL CONNECTION —

AFV)P/)b/754

POW

AUXILIARY GENERATOR (INSTALLATION)

FIG. 26.

a flexible exhaust pipe with a silencer that could be pushed out of the pannier door. It plugged into a junction box behind the front gunner's escape hatches. The generator had electric as well as pull start, and its ignition was switched on from the battery compartment. There is a socket in the battery compartment to enable a slave lead to charge the batteries as well.

On the floor behind the driver is the all-important toolkit.

There was a small fan in the roof of the driving compartment, next to the Besa (omitted from Mark I tanks with the 3in howitzer), intended – but generally described as inadequate – to draw fumes from the gun out. Both driver and front gunner had a 'distribution box' with a drop lead to connect their headsets and microphones to the Wireless Set 19 and enable them to talk to each other and the rest of the crew. The driver's box (distribution box No 1) had a button marked 'Call Commander', which activated a buzzer to attract the attention of the commander. The front gunner's box (distribution box No 2) had no buzzer. Both could only work on intercom and not transmit.

Fighting compartment

This section of the tank comprises the turret and, suspended from three struts below it, a floor. The floor is circular, with a diameter of

53.25in, and hangs on these struts from the turret base. It is kept centred by a hole in the floor that revolves around the drum housing a rotary base junction, which supplies power to the turret and connects the driver and front gunner to the intercom. Three crew members fit into this small area: the loader/operator, the gunner and the commander. Looking down from above, the loader/operator stands to the right of the main gun. He is responsible for operating the Wireless Set. He is also in

LEFT The turret floor of a Churchill Mark VII. It is very difficult to give a sense of how small a turret is by simply looking at a photograph. Here you can see, from the left, the commander's pedestal – the round stool (missing the plate for the centre, which he actually stood on), then the gunner's adjustable seat. At the foot of the gunner's seat are two pedals: the one nearest the seat adjusts its height; the pedal partially hidden behind the ammunition bin is the one that fires the guns (this particular type is operated by pressing down on it with the heel). The other kind is pushed by the front of the boot under the toes. The loader stands on the side nearest to the camera just behind the 'ready to use' ammunition stored on the turret floor. This ammunition bin has thin armour around the sides. Beyond the gunner's seat is his leg-guard with a strengthening bar across it. (*Tank Museum*)

LEFT The crowded turret of a Mark IV that has been converted from a 6-pounder gun tank to a 75mm version. (*Tank Museum*)

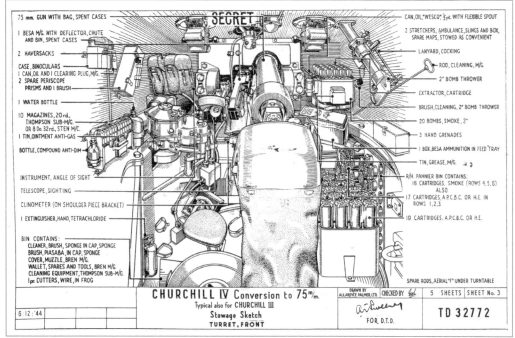

75 mm. GUN WITH BAG, SPENT CASES — SECRET

I BESA M/G. WITH DEFLECTOR, CHUTE AND BIN, SPENT CASES

2 HAVERSACKS

CASE, BINOCULARS
I CAN, OIL AND I CLEARING PLUG, M/G.
2 SPARE PERISCOPE PRISMS AND I BRUSH

I WATER BOTTLE

10 MAGAZINES, 20 rd., THOMPSON SUB-M/G. OR 8 Do. 32 rd., STEN M/G.
I TIN, OINTMENT ANTI-GAS

BOTTLE, COMPOUND ANTI-DIM

INSTRUMENT, ANGLE OF SIGHT

TELESCOPE, SIGHTING

CLINOMETER (ON SHOULDER PIECE BRACKET)

I EXTINGUISHER, HAND, TETRACHLORIDE

BIN CONTAINS:
CLEANER, BRUSH, SPONGE IN CAP, SPONGE BRUSH, PIASABA, IN CAP, SPONGE COVER, MUZZLE, BREN M/G. WALLET, SPARES AND TOOLS, BREN M/G. CLEANING EQUIPMENT, THOMPSON SUB-M/G. I pr. CUTTERS, WIRE, IN FROG

CAN, OIL, "WESCO", ½ pt. WITH FLEXIBLE SPOUT
2 STRETCHERS, AMBULANCE, SLINGS AND BOX, SPARE MAPS, STOWED AS CONVENIENT
LANYARD, COCKING
ROD, CLEANING, M/G.
2" BOMB THROWER
EXTRACTOR, CARTRIDGE
BRUSH, CLEANING, 2" BOMB THROWER
20 BOMBS, SMOKE, 2"
3 HAND GRENADES
I BOX, BESA AMMUNITION IN FEED TRAY
TIN, GREASE, M/G.
R/H. PANNIER BIN CONTAINS:
16 CARTRIDGES, SMOKE (ROWS 4,5,6)
ALSO
17 CARTRIDGES, A.P.C.B.C. OR H.E. IN ROWS 1,2,3
19 CARTRIDGES, A.P.C.B.C. OR H.E.

SPARE RODS, AERIAL "F" UNDER TURNTABLE

CHURCHILL IV Conversion to 75ᵐ/ₘ.
Typical also for CHURCHILL III
Stowage Sketch
TURRET, FRONT

DRAWN BY ALLARDYCE PALMER LTD. CHECKED BY 5 SHEETS | SHEET No. 3

6·12·'44 FOR D.T.D. TD 32772

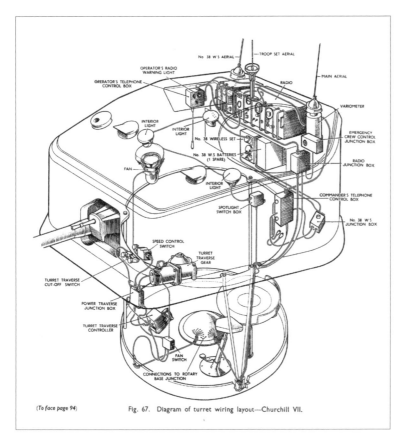

(To face page 94) Fig. 67. Diagram of turret wiring layout—Churchill VII.

the loader/operator in the turret and comprised only smoke ammunition – even though HE and illuminating flare rounds were available for infantry 2in mortars.

The all-important radio, the Wireless Set No 19, is stowed across the rear of the turret. The loader/operator has a junction box for his headset and microphone. Below the radio is a stowage shelf with compartments for circular 100-round Bren-gun drum magazines, together with other items and, later, the WS38 AFV.

Beside the loader/operator's legs is the battery compartment, containing two large 6V batteries in wooden cases, wired in series to produce 12V. This compartment also houses the voltage regulator for the battery-charging dynamo, controls for the petrol generator and the battery master switch. As part of the routine to start the tank, the loader/operator has to switch on a battery master switch by his side, and then turn round and crouch down to face the bulkhead behind him, to prime the carburettors and use the 'Ki-gass' pump to inject petrol directly into the inlet manifolds to help cold starting. This is not a tank for one person to operate from the driver's seat!

At the back of the fighting compartment are two interconnected CO_2 fire extinguishers, which are piped into the engine compartment.

To the left of the main gun are the gunner

ABOVE This illustration shows the turret of a Mark VII tank with the WS19 and WS38 installed. This is an early installation and the WS38 is running from its own dry battery rather than the tank's 12V supply. (*Author's collection*)

charge of loading the main gun and providing the ammunition boxes for the coaxial Besa as well as firing the 'bomb-thrower', the 2in mortar mounted in the turret roof. The bomb-thrower ammunition was stowed immediately in front of

RIGHT This is a view of the back of a Mark IV turret. The perspective of the viewer is as if he was standing where the gun is, and looking back. (*Tank Museum*)

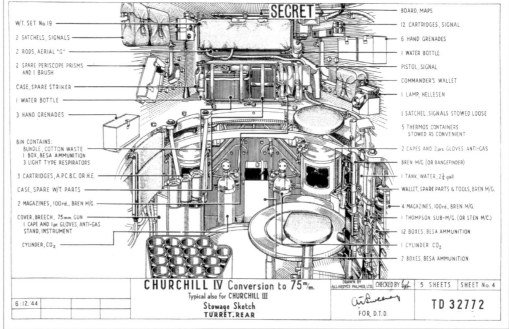

and commander. The gunner sits forward of the commander on a spring-loaded seat that adjusts for height. Later versions of the gunner's seat had a detachable backrest. On tanks with free elevation, the gunner has his right shoulder in a shoulder-piece that allows him to push the gun up and down. The gun cradle is clamped to the turret roof when elevation is not required, and this accounts for the consistent, slightly elevated, poise of the gun in many photographs. Those tanks with geared elevation had a handwheel by the gunner's right hand to elevate and depress the main gun and coaxial Besa. The Besa is close to the gunner's sighting telescope, just to his right. Firing was accomplished by a variety of means. In free-elevation gun tanks, there were two triggers below the main guns, one for the coaxial Besa and the other for the main gun. With geared elevation came a foot pedal to fire the guns. Later, electric solenoid firing became available as a supplement to the foot pedals.

On the gunner's left is a leg shield, a large plate to prevent his knees or legs being caught as the turret traverses. On this plate are mounted various junction boxes, a fire extinguisher and, most importantly, the traverse control box. This box has a handle that pivots right and left and operates the electric traverse motor. Rather cleverly, it does this not by acting as a giant variable resistor, but instead carries relatively low current and operates by changing the output from a dedicated dynamo in the gearbox compartment, which in turn energises the traverse motor to move slowly or quickly to right or left. An additional bonus of this system was that when the traverse control handle was released, the electric motor acted as a brake to stop the turret continuing its rotation past a target. Of course, this system means that electric traverse is only available when the engine is running. At other times, the gunner used a handle on the traverse mechanism to turn the turret. The gunner had a sighting telescope with a graticule that enabled him to aim both the main gun and the coaxial Besa, which was graduated to show the elevation needed for the different ballistic characteristics of the AP, HE and smoke rounds. On 75mm tanks, given the gun's ability to act as a form of artillery, the gunner had a ranging drum with a small spirit level that enabled him to calculate the elevation needed for indirect

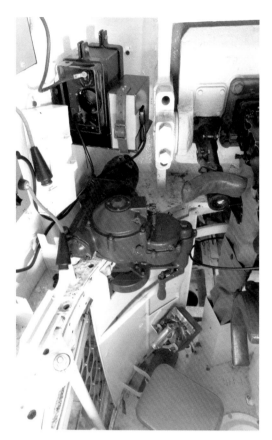

BELOW This very cluttered scene is the gunner's position (he sits on the small brown seat in the middle of the picture) in a Mark VII. This is the inside of the Mark VII in Bovington Tank Museum. *(Tank Museum)*

LEFT This is the gunner's position in a Mark IV. The WS38 is in the upper left of the photograph and the drop leads for the gunner's and commander's headsets can be seen. The green traverse gearbox is below, which allows the gunner to traverse manually. Below that are the electric traverse controller and the gunner's seat. To the right is the shoulder piece for the gunner to control the elevation of the gun and in front of it is the green canvas chute that takes spent cartridges from the Besa into a bin on the turret floor. The gunner's sighting telescope and brow pad can be seen next to the Besa. *(Author's collection)*

Seeing out of the tank was a matter of life and death for the crew. Once closed down, there were only three ways to view the outside world – through a periscope or, for the gunner and front gunner, through a sighting telescope. The driver, if it was safe to do so, could look through his armoured glass vision block.

The commander had a rotating cupola – a ring with two hatches (through which he and the gunner would also have to enter and leave). Early cupolas had two periscopes, one front and one back giving the commander a very limited field of vision. He had to rotate the cupola to change his field of view.

Later in the war the 'all-round vision cupola' was fitted. This was taller and had eight small periscopes (correctly called 'episcopes') that provided a relatively complete 360-degree view from the safety of the turret. Vision for tank commanders was a constant source of concern; if they put their head and shoulders out of the tank, they saw and heard much more. However, they were vulnerable to snipers and to shell splinters. Late in the war, some commanders took to having a plate welded behind their cupola to provide a shield against snipers.

The new cupola did cause some confusion; at least one tank (from 141st Regiment RAC, the Buffs) was knocked out by a British M10 tank destroyer whose crew were unfamiliar with the shape of the new cupola – which was similar to German designs – and mistook the Churchill for an enemy tank in the early morning mist in Normandy.

The gunner and front gunner each had a telescope with somewhat limited magnification, and with markings to help sight the main gun and the Besa machine guns according to range.

The type of sighting telescope varied with the mark of tank and calibre of gun. Marks I and II with 2-pounder guns used Telescope No 30 Mark I and IA, or No 33. Those with 6-pounder guns used the Telescope No 39. These telescopes gave a magnification of approximately 1.8 times and a field of view of some 21 degrees. On the Nos 30 and 33 telescopes, the cross wires were thick enough to obscure at least 2ft of the aiming mark at 1,000yd. On the No 39 telescope the thickness was halved. With the advent of the 75mm gun, the Telescope No 50 was issued, which had either 1.9 or 3 times magnification depending on mark and could be illuminated (the illuminated version was No 50 x 3 L Mark 1).

BELOW This is the periscope in the early version of the commander's cupola (it is seen here in the turret of a Mark IV, but it was used in all marks up to and including early Mark VII tanks). The two handles enable the commander to pull the cupola round. It runs on rollers (one can be seen, rusty, just to the left of the periscope). *(Author's collection)*

BELOW A tank commander in a Mark IV, V or VI with a plate welded behind the cupola to give protection against snipers. He is using the cupola flaps as additional protection. The blade vane sight is on the right of the picture and the round cap with four bolts protruding is the PLM mount for the Vickers K guns. *(Tank Museum)*

firing (in other words firing when he could not see the target, but was given a distance to it and provided with corrections to bring him on to target). In the Mark VII, there was also a very accurate traverse indicator, a clock-like dial to the gunner's left that gave precise readings for traverse to help correct indirect firing. On the Marks I through III and in Marks VII and VIII, the gunner also had a periscope which, like the loader/operator's, had a handle through the body of the periscope to help turn and angle it. The Marks IV, V and VI with their cast turrets did not provide for a gunner's periscope, save on the NA 75 where the Sherman gun required a 'peritelescope' for which a hole was cut into the turret above the gunner.

Behind the gunner is the commander. He has a pedestal to stand on, which can be adjusted for height, and a small circular fold-down seat for use when the cupola hatches are open. (On some Mark VII tanks this circular seat was replaced with a more comfortable one.) The turret crew have access to the intercom on the WS19, and the commander and loader are able to transmit on the 'A' and 'B' Sets. There was also a slightly different set-up for tanks with the WS38, used for radio communication with the infantry.

Radios and communications in the Churchill

In the Churchill tank the principal radio, or wireless set as it was then known, was the Wireless Set (WS) No 19.

The set came in three main variants, Marks I, II and III, and can be thought of as comprising four separate components:

■ The 'A' Set, a high-frequency transmitter and receiver for longer-range communications by voice or Morse code. It had a theoretical range – when using standard aerials – of some 10 miles for voice and 15 miles for Morse transmissions. It operated with frequency ranges of 2.5–6.25 megahertz (MHz) for the Mark I set, and 2–8 MHz for the Marks II and III sets.
■ The 'B' Set, a very high frequency (VHF) set for short-range communications with infantry and, in theory, between nearby tanks. Although said to have a line-of-sight range of up to a mile, in practice it appears to have

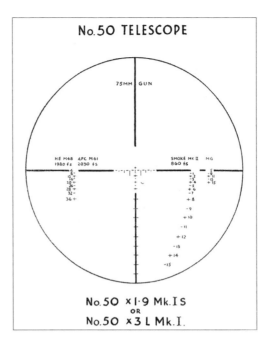

No.50 TELESCOPE

75MM GUN

HE M48 APC M61
1980 f.s 2030 f.s

SMOKE MK II   M G
860 f.s

No.50 × 1·9 Mk. I S
OR
No.50 × 3 L Mk. I.

had a significantly shorter range of a few hundred yards and to have been unreliable as a means of communication. It operated in the frequency range of 229–41MHz.
■ The intercom, enabling the crew to talk to each other.
■ The power supply to convert the 12V from the tank's batteries into the high and low voltages required for the valves. Current consumption varied according to what facilities were in use. Marks I and II used a maximum of 11.6 Amps if both 'A' and 'B' Sets were transmitting and the intercom was on. The equivalent figure for the Mark III set was 10.7 Amps. However, the sets drew a considerable current on start-up.

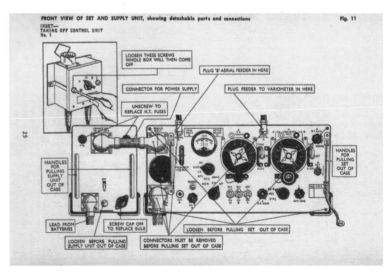

RIGHT The headset
and microphone for
the turret crew, with
'snatch plug'. *(Author's
collection)*

FAR RIGHT The
headset and
microphone used by
the driver and front
gunner. These two
crew members used a
different microphone
from the turret crew.
(Author's collection)

Each member of the crew had a headset and microphone with a rubber 'snatch plug' on a lead. This plugged into a 'drop lead' with a similar rubber plug, providing a reasonably waterproof connection that would pull apart quickly and easily if they had to leave the tank in a hurry (hence the term 'snatch').

For the technically minded, the output power of the 'A' set was up to 3W for voice, and up to 5W for Morse. The 'B' set had an output power of 0.4W.

The 'A' set used a tuning device for its aerial known as a variometer, which was attached to the turret roof directly below the 'A' set aerial;

this was generally an 8ft vertical rod. The 'B' set used a short 25in vertical rod sitting on top of a short pillar, with a protective cage around its base.

This remarkable radio was manufactured in England, Canada and the United States.

Some of the sets were produced under Lend-Lease contracts for Russian use. These sets have both English and Russian lettering on the front panels of the power-supply unit and the radio. A number of tanks also carried a smaller infantry radio, the WS No 38, to enable them to talk to the infantry that they were supporting, and these tanks can be distinguished by a third aerial on the turret. On early tanks this aerial was above the gunner, in front of the commander's cupola. Later, it was fitted to the rear of the turret roof.

In its original guise, the WS No 38 was a man-portable transmitter and receiver developed in 1942 to provide short-range communication for infantry. These sets operated on a frequency range between 7.4 and 9.2MHz. Their power output was a tiny 0.2W, used for voice transmission only. Their range was at best a 1 mile line of sight, using a 12ft-long rod aerial. This set, complete with its battery, was used in the Churchill with an 8ft aerial until the development of the Wireless Set No 38 AFV in 1944. This latter set worked off the vehicle

BELOW A Canadian-
produced WS19 Mark
III, produced under
the Lend–Lease
programme, showing
Russian Cyrillic script
on the labels. *(Author's
collection)*

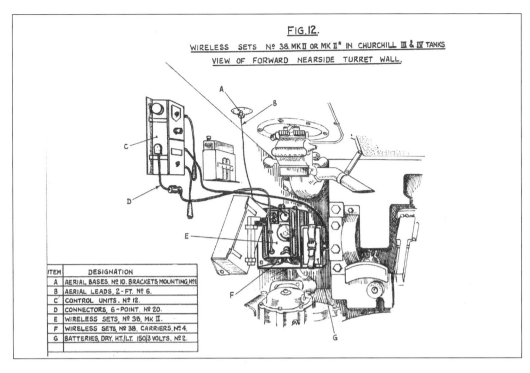

FIG. 12.

WIRELESS SETS Nº 38. MK II OR MK II* IN CHURCHILL III & IV TANKS

VIEW OF FORWARD NEARSIDE TURRET WALL.

ITEM	DESIGNATION
A	AERIAL BASES, Nº 10. BRACKETS MOUNTING Nº1
B	AERIAL LEADS, 2-FT. Nº 6.
C	CONTROL UNITS, Nº 12.
D	CONNECTORS, 6-POINT. Nº 20.
E	WIRELESS SETS, Nº 38, MK II.
F	WIRELESS SETS, Nº 38, CARRIERS, Nº 4.
G	BATTERIES, DRY, H.T./L.T. 150/3 VOLTS. Nº 2.

LEFT The installation of the WS38 in early Churchills. Note the lead (B) going to the hole in the turret roof for the aerial. *(Author's collection)*

batteries and was more easily connected to the crew radio and intercom systems.

In the Mark VII tanks an additional 'emergency crew control' was installed for the commander, comprising a small box with a socket and microphone so that the commander could speak to the crew even if the WS19 was damaged.

On the back of some tanks there was an infantry telephone in a box with a hinged lid, which was linked to the intercom. This originally used a headset and buzzer similar to that provided for the driver. Later tanks had a telephone-type handset. Before the

LEFT The emergency control transformer box and the Tannoy microphone that enabled the commander to use the intercom even if the WS19 was broken. This is the same type of microphone used by the driver and front gunner. *(Author's collection)*

FAR LEFT A crewman using the 'telephone' on the back of a Churchill ARV. Note that this utilises a Box, Distribution No 1 (the same as the one used by the driver) and that he is using the same type of microphone and headset as the driver. He is plugging these into the 'drop lead' of the distribution box to enable him to talk to the crew. The tank has a jettison fuel tank on the rear. *(Tank Museum)*

LEFT This photograph shows the same distribution box, but with a later 'telephone' type of handset. It is seen here fitted to the back of a Churchill Mark IV. *(Author's collection)*

.0 (Fig) ACCEPTED	G ACTIVITY	17 ADVANCE(D) (ING)	AIRBORNE	32 AIRCRAFT	F AIRFIELD LANDING GROUND	5 AMBER	58 APPROX AREA	A ARMD CARS	73 AIR BURST	P ATOMIC STRIKE	89 ARTILLERY
00 ATTACK(ED)(ING)	09 AXIS	N BATTALION	24 BATTERY	33 BLACK	41 BLOCKED	M BLUE	T BOWLINE	SWITCH ON	80 BOMBED HEAVILY AT	80 BOMBER	9 BOMBLINE
A BOMBS	SWITCH OFF	18 BOUNDARY	U BRIDGE(S)	B BRIDGE(S) DESTROYED AT	42 BRIDGE INTACT	50 BRIGADE	CAMOUFLAGED	66 CANAL	74 CARRY OUT	Q CALL SIGN	W CASUALTY(IES)
01 CENTRE LINE	H CHANGE(D)	CHANGE FREQUENCY TO...AT...HRS	25 CLOSE DOWN AT...HRS	35 COLUMN	G CONCENTRATE (D)(ION)	51 CONFIRM(ED)	59 CONTACT(ED)	SWITCH ON	I CONTROL	81 CORPS	90 COUNTER ATTACK(ED)(ING)
02 COVER	1 (Fig) DAMAGE(D)	SWITCH ON	26 DAY(S)	34 DEFEND (CE) (DED)	43 DEMOLITIONS	DEMOLITION AT.. ROAD IMPASSABLE	V DESTROY(ED)	67 DESTRUCTION	I DIGGING	82 DIRECTION (ING)(ED)	91 DISPERSED
D DISPLAY *	10 DIVISION	O DRO PH PRINTS AT..	V DUMMY(IES)	DUMP(S)	H EAST	N ENEMY	6 ENGAGE(D) (ING)	68 ETA	H EXPLOIT(ING)	FDL	SWITCH ON
03 FACING	11 FIGHTER(S)	19 FIGHTER ESCORT	W FIRST LIGHT	36 FLAK	44 FLARE(S)	53 FOG	W FWD TPS REACHED	69 FREQUENCY (IES)	75 'G' AIR BRANCH	83 GLIDER(S)	92 GREEN
C GRIDDED OBLIQUES	I GUNS	2 GRID REF	27 GUNS A TK	37 GROUND BURST	SWITCH ON	60 GUNS FIRING	C GUNS SP	K HQ	76 HARBOUR(ED)	84 HEAVY	X HELD UP GENERAL LINE
04 HOURS	12 HELICOPTER	20 IMMEDIATE (LY)	X INFANTRY	INTENTION	P INTERPRET-ATION	54 JUNCTION	61 LANDMARK	D LAST LIGHT	76 LEFT	LINE	93 LOCATION
05 LORRIED INF	13 LOW CLOUD	P MT	28 MAP REF	MESSAGE DROPPING STA	J MILES	MINE FIELD	SWITCH ON	E MORTAR(S)	MOVE(D)(ING)	85 MOVEMENT	MY LOCATION IS
06 NEUTRALIS-ATION	J NIGHT	O NO MOVEMENT SEEN	SWITCH OFF	39 NORTH	NUCLEAR	Y OBJECTIVE/TARGET	X OBLIQUE	7 OBSERVE	M OPEN AT.. HRS	86 OPS	94 OWN TPS
D PARACHUTE (ISTS)	K PHOTOGRAPH (IC)	21 POINT	29 POSITION	POSITION FWD TPS IN 2 HRS	55 POSSIBLE	PRIORITY	62 PROBABLE	68 RV	77 RAILWAY	T RANGING	95 REAR
SWITCH OFF	14 REAR LINK	22 RECCE	Y RECOGNITION SIGNAL	4 RED	K RECEIVE(D)	56 REFUSED	63 REGISTRATION	70 REPORT(ED)	78 REPORT POSN LEADING TPS	87 REQUEST(ED)	Z RESISTANCE
E RESULT	15 RIGHT	R RIVER	3 ROAD	40 SCI	47 SCALE	R SEND	SERVICE(ABLE) /PASSABLE	71 SIGNAL(S)	N SITREP	U SMOKE	96 SORTIE
07 SOUTH	L SQUADRON	23 STATIONARY	Z SS FORM	SWITCH ON	48 STRONG	57 SUCCESSFUL (LY)	64 SUITABLE	F SUPPORT	88 SWEEP	TAC R	97 TANK(S)
08 TENTACLE	16 TIME CHANGED TO	S TOWN	30 UNABLE TO CARRY OUT TASK REF'D TO	TRACKED VEHS	UNTRACKED VEHS	S URGENT	65 VERTICAL	VILLAGE	8 WEATHER PREVENTS	SWITCH OFF	98
F WHITE	M WILL ADVANCE TO...AT..HRS	T WILL ATTACK AT...HRS	31 WILL WITHDRAW AT...HRS	E WIRELESS/RADIO SILENCE FROM	49 WITHDRAW-ING	SWITCH OFF	WOOD(S)	72 X ROADS	YARD(S)	V YELLOW	99 YOUR

ASSU No I

SERIES B		ASSU No I	(W.O. Code No. 14964)	CARD 5

ABOVE An example of a Slidex card which, when laid in a holder with a sliding cursor along the top and side, enabled messages to be sent in a simple code. The exercise was time consuming but provided at least a short-term measure of security. The code settings were an item that the crew was to remove if they abandoned a tank. *(Author's collection)*

intercom system, there was a 'gong' button on the back of the tank to attract the crew's attention and communication was carried out by the commander putting his head out and – on some occasions – throwing down a spare headset and microphone on an extension cable. Infantrymen were also known to hammer on the sides of the tank to attract a crew's attention. As can be imagined, with the limited visibility

DER FEIND HÖRT MIT
(The enemy listens too)

Seven Steps to Slidex

1. Prepare your message—be brief.
2. Adapt it, as far as possible, to the language of the card to be used.
3. Consider the result from the enemy point of view.
4. Decide which parts are dangerous. If possible place these parts together.
5. Encode only these dangerous parts of the message.
6. Consider the whole message again as it will reach the listening enemy.
7. If in doubt—use S L I D E X.

REMEMBER THAT INSECURE MESSAGES CAN HAVE DISASTROUS RESULTS. YOU MUST BE SECURE.

The War Office
(Signals 9)
June, 1944

RIGHT Slidex instructions, explaining the need for encoding. The German Army had very capable radio interception techniques and, particularly during the early war period, it had learned a great deal from Allied radio communications. *(Author's collection)*

available to the tank crew, it was a hazardous business standing next to the tank when it might manoeuvre at any point.

As well as being able to use the radio, with its associated voice procedures, the loader/ operator had to be able to send and receive Morse and to encode messages using the Slidex code system. This was one of the items that had to be removed if a tank was abandoned, to avoid it being captured and giving away the current code settings.

A word of warning on restoration of radios

Wireless set restoration presents a number of hazards that are beyond the scope of this book. Some of the lettering on the front panels is luminous, and potentially therefore used radioactive paint. Secondly, the sets use high voltages and it is extremely risky to power up a set that has not been competently checked. Among other risks are exploding capacitors, aside from any electrical shock dangers. Lastly, it is important to check the applicable rules and laws on wireless telegraphy; depending on jurisdiction, these may well make it illegal to own or use the sets on some or all of their frequencies and also require the user to be licensed before operating them.

Engine

This is a 12-cylinder, horizontally opposed petrol engine. It develops a nominal 350bhp at 2,200rpm and maintains torque of some 960ft/ lb all the way from 800rpm to 1,600rpm. Here are the bare facts and figures:

- Weight when dry: 3,376lb.
- Number of cylinders: 12.
- Cylinder bore: 5in.
- Stroke: 5.5in.
- Compression ratio: 5.5 to 1.
- Capacity: 1,296cu in (just over 21 litres).
- Exhaust valves: sodium cooled with stellite-inserted valve seats.
- Pistons: light aluminium alloy.
- Spark plugs: two 18mm thread per cylinder with screened ignition harness.
- Ignition coils: four 6V coils in series with 6V ballast resistors.
- Nominal governed speed: 2,200rpm.

- Carburettors: four 46mm Solex Model 46 F.N.H.E.
- Lubrication: dry sump with de-aerating and oil cooling systems.
- Air cleaners: two AC two-stage, centrifugal and oil bath types (early tanks had a single-stage cleaner).

The cylinders are arranged in two blocks of six, and each block has half of the crankcase cast in. The right-hand half carries the crankshaft in seven main bearings. The crankshaft has six throws and the connecting rods from each opposing pair of pistons operate on one crankpin. To achieve this, the left-hand side of the crankcase is set slightly forward of the right-hand block. In most engines there are cast-iron dry liners, 3in long, in the top end of the bores. Some engines appear to have been rebuilt with full-length liners.

The four cylinder heads are made from high-tensile molybdenum cast iron with internal water passages in the castings.

The flywheel is cast-iron alloy with the rear face acting as the friction face for the single-plate clutch.

The piston connecting rods are made from nickel chrome steel, drilled throughout their length for oil to feed the big-end bearings, small-end bearings and, through a hole at the top of the rod, for a jet of oil to cool the underside of the piston crown.

The pistons are made from aluminium alloy with five piston rings. As the crowns of the pistons are shaped for combustion control, and

ABOVE An engine undergoing rebuild. The flywheel has been attached but the cylinder heads have yet to be fitted. Note the sloping tops on the pistons, which are 'handed' for each side of the engine. *(Author's collection)*

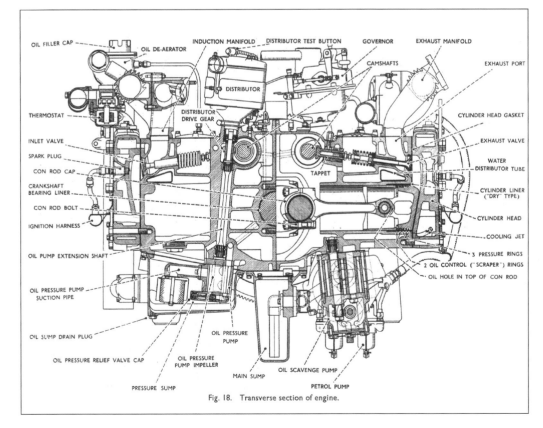

Fig. 18. Transverse section of engine.

LEFT A transverse section of the engine. *(Author's collection)*

also have a slot in the skirt on the non-thrust side, they are 'handed' and not interchangeable from left to right. Pistons recovered from engines being rebuilt will be found to have their nominal size or oversize stamped on the crown.

The engine has two camshafts, made of case-hardened nickel steel, running in four white-metal-lined shell bearings. Each camshaft drives one distributor, with one camshaft running the governor and the tachometer generator. There are two duplex roller chains to drive the camshafts set in a timing case cover at the front of the engine.

There are two valves per cylinder, sitting above the cylinders at a slight angle. Stellite seat inserts are used for the exhaust valves, and the heads and stems of these valves are sodium filled. This is important to know if you are machining the valves, as the sodium filling is combustible and should not be allowed to come into contact with the air.

Lubrication is carried out with a dry-sump system. There are two oil pumps, both gear-driven. One is the pressure pump and the other a scavenge pump. The system is pressure-relieved at approximately 50lb/sq in.

Four Solex 46 F.N.H.E horizontal carburettors are fitted, each serving three cylinders through its own inlet manifold. The throttle spindles of the carburettors are connected together by rods and levers, which are linked to the governor. Below the governed speed, the accelerator pedal operates a hydraulic cylinder that actuates the throttles of each carburettor. Care must be taken to adjust the connecting rods in this system, as they can bind and jam, preventing the engine revs from dropping correctly. A 'Ki-gass' priming pump is fitted at the back of the fighting compartment to facilitate cold starting; this delivers a spray of petrol directly into the inlet manifolds.

The clutch is a Borg & Beck design with a disc diameter of 18in. Pressure on to the driven plate is provided by 24 helical springs around the clutch cover in two staggered rows. The clutch is disengaged by six toggle levers that lift the pressure plate.

The ignition system comprises two distributors driven at half engine speed, incorporating automatic timing advance, and four 6V coils. The system also includes

a solenoid that, when the starter button is pressed, cuts out two resistance coils and allows 12V to go directly to two coils to provide a stronger spark for starting purposes. Each cylinder has two spark plugs. The originally correct ignition timing is 3 degrees before TDC (top dead centre), but with modern fuels and rebuilt engines that have skimmed heads, some experimentation on timing is advisable.

There is a belt-driven dynamo mounted on top of the engine for charging the two 6V batteries.

An Amal duplex mechanically driven fuel pump is fixed to the bottom of the engine, accessible through traps in the floor of the hull. There is a hand-priming pump handle in the back of the fighting compartment. This operates a cable connected to the diaphragms in the fuel pump and this manually lifts petrol to the float chambers for starting when cold, or when the petrol in the chambers has 'cooked off' – evaporated – on a hot engine. There is a selector lever next to the front gunner, which sets a valve under the engine to draw petrol from left-hand or right-hand tanks. The auxiliary generator has an electric SU diaphragm pump of its own.

The cooling system is duplicated on each side of the engine, so that each bank of six cylinders has its own self-contained circuit of water pump, radiators, thermostats and temperature gauge. The cooling system holds 26gal, 13 on each side, with four thermostats that start to open at 175 degrees Fahrenheit, and are fully open at 190 degrees.

The engines went through a variety of modifications and improvements, which are designated by the addition of a suffix to the engine number. The original modification was to increase the oil feed to the distributor and introduce 37mm carburettors, and this resulted in the suffix 'R'.

After that further changes were designated as follows:

- R1: positively driven Amal fuel pump introduced; de-aerator improved.
- R2: connecting-rod big-end bolts introduced in place of studs.
- R3: mechanical, adjustable tappets, with modified camshaft and valves introduced (superseding hydraulic tappets).

- R4: crankshaft main centre bearing cap strengthened. 'Ki-gass' pipes changed from steel to copper.
- R5: connection for air compressor introduced. Timing chain tensioner bracket strengthened.
- R6: detail changes made to piston rings, crankcase oil retaining ring and distributor gears.

Gearbox and transmission

Gearbox – facts and figures:

Weight with compressor: 2,016lb approx.
Ratios for later gearbox:

> 1st – 120.2:1.
> 2nd – 42.32:1.
> 3rd – 23.75:1.
> 4th – 15.02:1.
> Reverse – 168.5:1.

Speed in each gear at 2,200rpm with later gearbox:

> 1st – 1.59mph.
> 2nd – 4.51mph.
> 3rd – 8.03mph.
> 4th – 12.7mph.
> Reverse – 1.13mph.

As well as containing the change speed gears, the gearbox incorporates the mechanism for turning by means of a controlled differential. The system uses epicyclic gear trains in conjunction with a differential gear in such a way that when a steering brake is applied, the output shafts of the differential are controlled – thereby reducing the speed of one track and speeding the other one up.

The tank can be made to pivot on the spot in neutral. This is, of course, a dangerous capability when the tank is in a confined space – inadvertently touching the steering controls with the engine running will cause it to slew and may injure or kill anyone standing alongside, as well as damage neighbouring vehicles, equipment or buildings.

The original, H4, gearboxes went through a series of modifications and, as with engines, these are indicated by suffixes to the gearbox number.

On the introduction of the H41 gearbox, changes were denoted by suffixes to the gearbox type designation.

ABOVE The gearbox installed in the Mark IV hull. On the left are the rods and levers which are the means of changing gear. They go all the way under the turret floor to the gear lever beside the driver. On the right, with a filler cap, is the compressor on top of the gearbox. The steering brakes have had the drums removed to reveal the friction linings. *(Author's collection)*

LEFT A Churchill gearbox laid bare. *(Author's collection)*

BELOW Gearbox internals explained. *(Author's collection)*

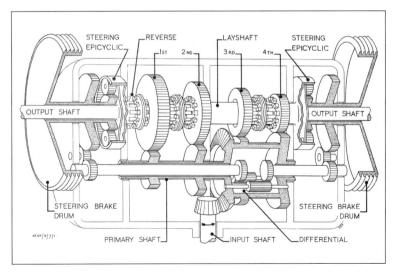

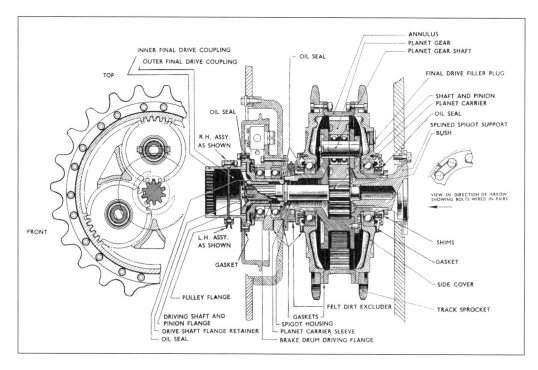

Fig. 101. Changing final drive unit.

Final drives – facts and figures:

- Weight when dry: 1,100lb.
- Oil capacity: 1.75gal.
- Teeth on sprocket: 23.

There are two final drives on the tank, on each side at the back, in line with the gearbox. These convert the output from the gearbox into the motion for the tracks.

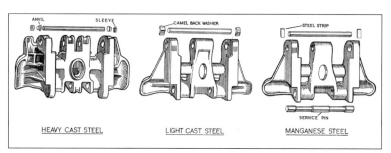

Front idlers – facts and figures:

At the front of the tank, on each side, are the front idlers which, like the final drives, each have 23 teeth and are passive drums on ball and roller bearings that serve to keep the track in alignment. Each idler weighs 674lb and adjusts forwards and backwards to maintain the correct track tension.

Track and suspension – facts and figures (for manganese steel track):

- Pitch of track links (or plates, as they were known): 7.96in.
- Number of links: 72 per side; additional links carried as spares.
- Weight of one link: heavy cast-steel track (Dieppe era), 59.25lb. Manganese steel track, 48lb.
- Number of suspension bogies: 22 in total, 11 on each side.

The ground pressure should be 14lb/sq in when the tank is fully laden, and on the later Mark VII gearbox, a speed of 12.7mph should be possible without wearing the track out too quickly (a figure of 2,000 miles of track life was expected).

The track was designed to be able to crush moderately large stones and keep out barbed wire. The links run dry, without lubrication. Each

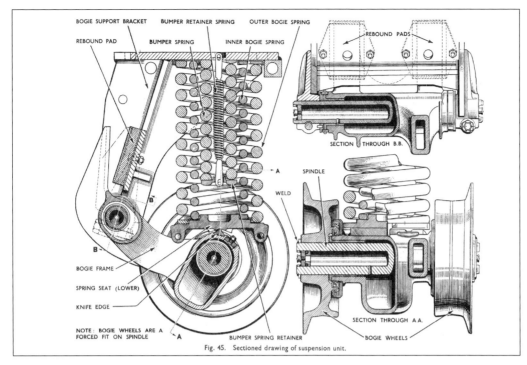

Fig. 45. Sectioned drawing of suspension unit.

Labels on figure: BOGIE SUPPORT BRACKET · BUMPER RETAINER SPRING · OUTER BOGIE SPRING · REBOUND PAD · BUMPER SPRING · INNER BOGIE SPRING · REBOUND PADS · SECTION THROUGH B.B. · A · B · B · SPINDLE · WELD · A · BOGIE FRAME · SPRING SEAT (LOWER) · KNIFE EDGE · SECTION THROUGH A.A. · NOTE : BOGIE WHEELS ARE A FORCED FIT ON SPINDLE · A · BUMPER SPRING RETAINER · BOGIE WHEELS

track pin is held in by a stainless-steel retaining bar welded to each side of the link, covering about half of the hole through which the pin is inserted. So that temporary track repairs can be carried out without welding equipment, a stepped 'service track pin' can be used without needing a retaining bar. These pins stay in place because they have grooves that correspond to the fingers on the track links. However, they are only a temporary replacement and should be replaced with a standard link as soon as possible.

The 22 bogie units – 11 on each side – have springs to absorb deflections caused by uneven ground.

Each bogie wheel is press-fitted on to the axle, which rotates in white metal bearings. The axle is hollow and acts as an oil reservoir. This is replenished through a grease nipple at each end of the axle.

Brakes – facts and figures:

There are two sets of brakes. One pair, on the gearbox, is for steering; the other pair, on the hull sides, is for stopping the tank.

■ Diameter of brake drums: 20in.
■ Width of linings: 3.5in.
■ Area of steering and stopping brakes: 123sq in.

■ Weight of brake assembly: 188lb each (of which the brake drum weighs 94lb).

Both sets of brakes are Lockheed hydraulic and each set consists of two shoes mounted on a back plate. The shoes are interchangeable. The drums have the same dimensions for both steering and braking; however, the steering drums are ribbed for cooling purposes, while the stopping brake drums are plain.

The operating cylinders and rocking beams in the stopping brakes are different from those in the steering brakes and are not interchangeable.

Between the pairs of brake shoes is a pair of Lockheed hydraulic cylinders, connected in series to ensure that their operation is synchronised. The brakes are all operated by hydraulic pressure. In the case of the stopping brakes, this pressure is applied by the footbrake and by the handbrake. While the handbrake will lock the brakes on, it cannot be relied on for parking on a slope, as the hydraulic pressure will fall once the engine has stopped and allow the brakes to release. When parking on a slope, first gear should be engaged. The steering brakes are operated by the steering handlebar, with a separate master cylinder for each of the right-hand and left-hand steering brakes. Air pressure assists the steering brakes.

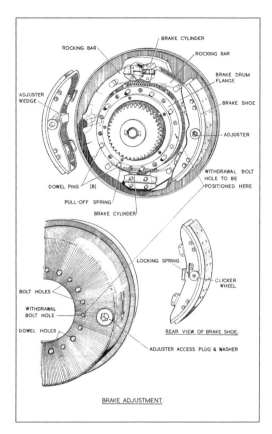

Crew stowage items

The crew had to be a self-sufficient group in battle, able to fight, carry out maintenance and some repairs, eat, sleep and also treat wounds. Their personal effects, rations, first-aid equipment and toolkit all had to fit in, or on, the tank.

The stowage diagrams show what was officially stowed and where. In fact the crews took to hanging more and more items outside the tank, often in ammunition boxes welded to the turret.

Basic dimensions of the Churchill

The dimensions of the tank should be very straightforward, but it is interesting to note that different sizes are quoted in official publications. These fluctuations are most notable when it comes to height and this can be accounted for by the fact that there is scope to measure height to different points, whether it is the top of the 'B' set aerial base, or the periscope housing. The following dimensions are taken from the tank handbooks and show surprisingly little variation:

Churchill Marks I and II

■ Overall length: 24ft 5in.

Air pressure at approximately 80lb/sq in is supplied by a two-cylinder Clayton Dewandre compressor mounted on top of the gearbox.

RIGHT AND
OPPOSITE Some of
the items of external
stowage on a Mark IV,
which has been
converted to take a
75mm gun in place of
the original 6-pounder.
(Tank Museum)

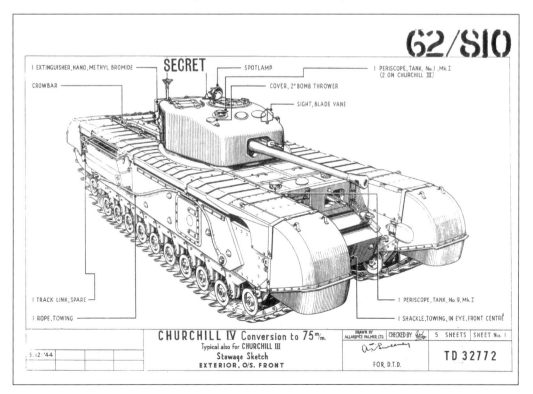

- Overall width with air louvres: 10ft 8in.
- Overall width without air louvres: 9ft 2in.
- Overall height: 8ft 2in.
- Length of track on ground: 12ft 6in.
- Width over tracks: 9ft 1in.
- Clearance under hull: 1ft 8in.
- Gross weight: 35/40 tons.

Churchill Marks III, IV, V and VI

- Overall length: 25ft 2in.
- Overall width with air louvres: 10ft 8in.
- Overall width without air louvres: 9ft 2in.
- Overall height: 8ft 2in.
- Length of track on ground: 12ft 6in.
- Width over tracks: 9ft 1in.
- Clearance under hull: 1ft 8in.
- Gross weight: 35/40 tons (some publications list 39 tons fully stowed).

Frontal armour:

- Hull: 3.5in maximum depending on location.
- Turret frontal armour: 4.5in maximum on Mark IV (which has a cast turret that varies in thickness across the frontal area) and 3.5in on the Mark III turret.
- Side: 2.5in maximum, without appliqué armour added.
  Appliqué armour adds 1.25in by the addition of a panel either side of the mantlet opening on the turret front (Mark III only) and 20mm to sides of hull and (on the Mark III only) sides of the turret.

Churchill Mark VII

As above, save for weight indicated as 41 tons (some publications state 40 tons).
Frontal armour:

- Hull: 6in maximum.
- Turret frontal armour, being cast, varies from 3.5in at one point to 6in in others.
- Side: 3.75in.

Capacities and consumption (approximate)

- Petrol: 150gal split equally between left and right tanks, and in Churchills with jettison tanks carried on the rear, 32.5gal in this tank.
- Petrol consumption: the manuals quote the not very useful figure of 0.57lb weight of fuel per bhp/hr. However, the report on Operation Trent notes that tests carried out with 25 reworked Churchill tanks showed an average of 0.68mpg for those tanks that did 80% road running and 0.56mpg for those tanks that did approximately equal road and cross-country running. Interestingly, the report notes that oil consumption in this test averaged 42.5mpg.
- Cooling system: 26gal (13 on each side).
- Oil in engine: 3gal.

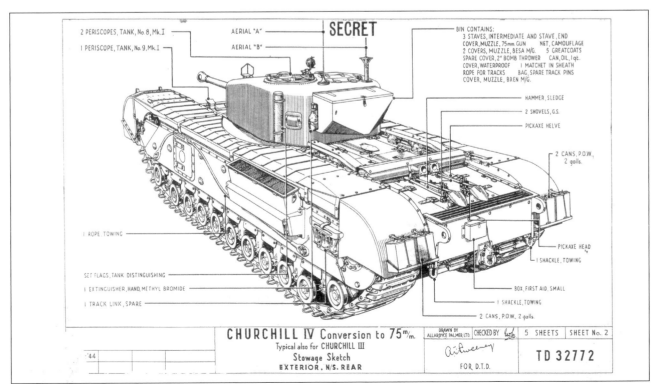

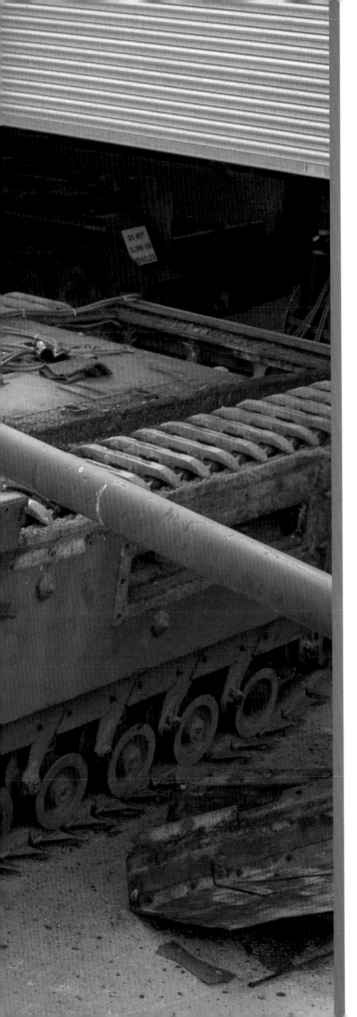

Chapter Five

Restoring the Churchill

⊏━(●)━━━━━━━━⊐

'Why restore a Churchill?' is a question sometimes asked. 'Look at its place in history' is the reply. Careful attention and real dedication are needed to bring this iconic tank back to life. Here are some of the issues faced in a Churchill restoration and how they were overcome.

OPPOSITE The Churchill Twin ARK/Mark IV hull with the bridging supports cut off, the turret blanking plate removed and the air intake louvres removed. Note the raised driver's and co-driver's periscopes and the very different post-war track guards. *(Author's collection)*

101

RESTORING THE CHURCHILL

ABOVE Winston
Churchill walks away
from a demonstration
of the tank that bears
his name. *(Author's
collection)*

Why restore a Churchill?

Of all the British tanks produced and
operated during the course of the Second
World War, the Churchill has a strong claim to
be the best and most enduring.

First, it saw front-line service for much of the
war, from the attack at Dieppe to the closing
stages of the campaigns in Italy and north-
west Europe. Churchill tanks also served in
North Africa, not only in Tunisia but also at El
Alamein. They then saw service in the Korean
War as a gun tank and Armoured Recovery
Vehicle (ARV) before being retired from duties
in a variety of specialist roles in the British Army
as late as the 1960s.

Second, the Churchill was tough and, although
far from immune to attack, tended to catch fire
less quickly, enabling the crew to escape more
easily than, say, the Sherman or Cromwell tanks.
This was partly attributable to armour thickness
and ammunition stowage, but also to the pannier
doors that allowed the driver and co-driver to
escape less obtrusively than if they had used the

top hatches. By the time of the Mark VII – but even
with the uparmoured earlier tanks – the Churchill
was remarkably able to survive hits, even those
that penetrated the armour.

Third, even with a relatively poor main gun,
it was successful in its primary role – that of
helping the infantry as they moved on the
battlefield. In doing so it earned the eternal
gratitude of many infantrymen. During most
of the war, that was the most important
requirement of an Allied tank: neither infantry
nor tanks alone can win and hold ground; only
a mixture of tanks and infantry (and artillery) can
do that successfully.

Last, there is the role of Winston Churchill
in the tank's evolution. Who would not see the
added attraction of preserving the history of a
tank linked to him?

This sentiment was, of course, lacking in the
era after the Korean War. By then, the Churchill
as a gun tank was redundant and in its place
the Centurion tank had proved its worth and
was to be a very successful tank for many years.

Apart from the few Churchills that remained

in service after Korea as engineer and specialist vehicles, the remaining Churchill tanks were consigned to firing ranges as targets and markers, and to the scrapyard.

It seems most unfortunate that a tank that had contributed so much to our history had been allowed to disappear without many examples being preserved, even in static form, let alone as running vehicles.

ABOVE Two Mark IV and a Mark I/II Churchills in a scrapyard after the war. *(Tank Museum)*

BELOW LEFT A derelict Churchill AVRE hull on a firing range. *(Author's collection)*

BELOW RIGHT A Mark IV with a 75mm gun on a firing range, filled and surrounded with concrete to prevent it from disintegrating with repeated hits from (in this case) 84mm Karl Gustav rounds. *(Author's collection)*

The tanks under restoration

LEFT The Twin ARK/Mark IV before removal of the bridging components. *(Author's collection)*

CENTRE This card shows the conversion details of various Churchills, as well as their original destinations. T31579 is the Churchill Project Mark IV, and it is tenth from the top in this list. As you can just make out, it had an engine change to engine number 15381R. *(Tank Museum)*

BELOW This is the contract allocation for T31579. The card illustrates one of the problems in trying to decipher precisely how many Churchills were built (see top right on the first card). Although the card states that the order was for Churchill I, II and III, ours of course is a Mark IV. It seems likely that it began as a Mark I or II and became a Mark IV when those types were declared obsolete. This order is one of the very first ones placed, dated 30 May 1941. *(Tank Museum)*

RIGHT This is the plate on the front of the Mark IV's engine, showing the number A22 (for Churchill)/15381/R, corresponding to the record card. The engine had been overhauled in April 1954. *(Author's collection)*

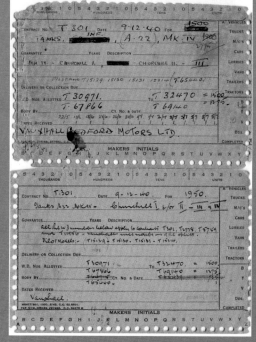

1 This is the engine compartment of the Mark IV when it was first opened.

2 Here is the engine in the Mark IV after the engine decks have been removed.

3 The engine removed from the tank …

4 … and the gearbox.

5 This is the condition of the driver's compartment once the ARK had been opened. To the right is the instrument panel. The vision port can be seen painted green (it would be too obvious as a target if painted silver like the rest of the inside was after the war). The box at the top right of the picture held a spare Triplex vision box (originally this was where the spotlight was stowed).

6 The ARK co-driver's compartment at the start of the restoration. The WS19 was stowed beside the co-driver, hence the junction boxes on the side of the hull. Note the co-driver has the extension of the steering tiller.

7 The hull is stripped out. Here the pedals and steering tillers can be seen.

8 The ARK driver's compartment has now been stripped bare.

9 This is the ARK gearbox compartment. The three large holes at the back allow the air drawn in through the louvres and the radiators, which circulates over the engine and gearbox, to exit. *(All photos on this page, Author's collection)*

Mark III pre-restoration

10 The engine and gearbox were still fitted to this Mark III. From the left:
the gearbox with top cover removed and compressor missing; the 'Sirocco' fan, which surrounds the clutch and flywheel; finally, the engine block with the remains of the two distributors visible. The small silver-coloured object to the left of the distributors is the rev counter drive.

11 This is the same Mark III with its hull stripped bare. Note the damage to the hull at the base of the photograph.

12 The Mark III turret seen from the inside. Note the damage caused by the shell or mortar that landed by the commander's cupola.

13 An external view of the Mark III turret showing damage.

14 This is the back of the Mark III during restoration. The back plate has been removed revealing the three holes through which warm air from the engine exits. It then hits the (removed) back plate and is forced up through slats on top, and also downwards. There is one surviving 4inch smoke discharger on the back, on the left side.

15 To detach the side armour the large bolts that hold it to the inner mild steel plate have to be cut. This bolt is just under the hole for the air louvre in the engine compartment.

16 The side armour coming off. Note the mild steel sheet behind. The appliqué armour can be seen clearly in this photograph.

17 The side armour coming off the other (left) side of the tank. The cut bolts can be seen projecting through the armour plate.

18 The Mark III hull after shot-blasting and priming. *(All photos on this page, Author's collection)*

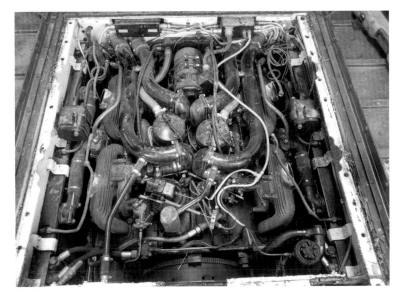

Challenges of restoring a Churchill

First of all, the engine. This is a side-valve engine; that is to say, the cylinder heads and valves are horizontal, forming the sides rather than the top of the engine. The engine is in effect a 'flat 12' comprising two six-cylinder blocks forming a shared crankcase. The crankshaft has six throws; in other words each opposing pair of pistons moves on the same part of the crankshaft. The flat construction of the engine had the advantage of keeping the profile of the tank lower, but it means that

ABOVE A view of the Churchill's engine, seen here from the gearbox end with the engine decks removed. Compare this view with the diagram below to identify each component.

BELOW The engine explained. *(Tank Museum)*

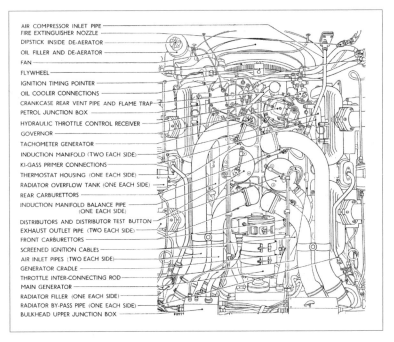

AIR COMPRESSOR INLET PIPE
FIRE EXTINGUISHER NOZZLE
DIPSTICK INSIDE DE-AERATOR
OIL FILLER AND DE-AERATOR
FAN
FLYWHEEL
IGNITION TIMING POINTER
OIL COOLER CONNECTIONS
CRANKCASE REAR VENT PIPE AND FLAME TRAP
PETROL JUNCTION BOX
HYDRAULIC THROTTLE CONTROL RECEIVER
GOVERNOR
TACHOMETER GENERATOR
INDUCTION MANIFOLD (TWO EACH SIDE)
KI-GASS PRIMER CONNECTIONS
THERMOSTAT HOUSING (ONE EACH SIDE)
RADIATOR OVERFLOW TANK (ONE EACH SIDE)
REAR CARBURETTORS
INDUCTION MANIFOLD BALANCE PIPE
            (ONE EACH SIDE)
DISTRIBUTORS AND DISTRIBUTOR TEST BUTTON
EXHAUST OUTLET PIPE (TWO EACH SIDE)
FRONT CARBURETTORS
SCREENED IGNITION CABLES
AIR INLET PIPES (TWO EACH SIDE)
GENERATOR CRADLE
THROTTLE INTER-CONNECTING ROD
MAIN GENERATOR
RADIATOR FILLER (ONE EACH SIDE)
RADIATOR BY-PASS PIPE (ONE EACH SIDE)
BULKHEAD UPPER JUNCTION BOX

some aspects of restoration and maintenance are more difficult. Changing spark plugs, for example, is not possible on a hot engine – it involves reaching down in a narrow space between the engine block and the radiators, close to the exhaust manifolds.

These exhaust manifolds, four in all, rise from the top of the engine and are the reason that all of the engines the Churchill Tank Project members found were seized – even those that had never been used, but had been stored outside. The explanation is simple enough: the exhaust system was generally open to the elements and water then travelled down the exhausts, through those exhaust valves that were open, into the cylinders. In most cases this led to piston rings rusting in the bores and the pistons being well and truly stuck. For those engines that had been outside, water in the block sometimes led to frost cracking of cylinder walls. This can be repaired, not least because the cylinders are lined, and new liners can be used after the crack is repaired.

The engine peripherals are mostly on the top of the engine: four carburettors, two distributors, a governor and the hydraulic throttle, not to mention air intake and exhaust pipes, rev counter drive and a dynamo. Because they sit on top of the block they are very prone to corrosion and other damage. The linkages from the throttle to the carburettors are almost bound to need rebuilding, and the carburettors need to be completely dismantled

and overhauled. The governor is a delicate mechanism that often rusts if water gets in and should be dismantled and checked. The distributors are also very easy to damage – particularly in removing the distributor cap and rotor arm, both of which are made from a plastic that seems to become more brittle with age. If necessary, these can be replaced with modern items made by 3D printing. The rev counter drive, a small dynamo, generally seems to survive years of neglect and needs only a clean and rebuild. The exhaust manifolds are generally sound, although some of ours had small holes in them; however, the air pipes that connect the four carburettors to the air filters are predisposed to corrosion and will probably need to be repaired. Of course, all gaskets and seals need to be replaced and, with the advent of laser cutting, this has become much easier to organise, although they can also be made the 'old-fashioned' way from gasket paper and other suitable materials. It should be noted that at least some of the original gaskets probably contain asbestos.

The engines comprise two blocks, two sumps and four cylinder heads. After removing all external ancillaries and taking the cylinder heads and sumps off, the blocks need to be pulled apart carefully to enable work on the internals. It is very easy to break pistons in the process of separating the blocks, and great care and patience is recommended, generally using two very large hydraulic jacks, with beams across the cylinder blocks bolted into the holes for the head bolts.

Care also needs to be taken in handling and dismantling the engines as there is an asbestos heat shield on top, in addition to possible asbestos in the gaskets mentioned above. One unexpected problem that we found was that many spark plugs had rusted solidly into the cylinder heads and had to be very carefully drilled out. The heads themselves, although seemingly massive, are prone to warping, and should be checked and almost certainly skimmed. New head gaskets will be needed, and careful and repeated checking and tightening of the head bolts is required until the system has bedded down.

Under the engine are two sumps; some examples are made from steel, others from

BELOW This is the engine block being split. The reason that the lower part is painted red is that the sumps sat under this part of the engine block.
(All photos on these pages Author's collection)

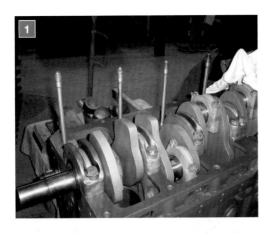

1 The crankshaft in place.

2 Pistons that have been removed from old engines await cleaning and checking.

3 The dynamo is rebuilt.

4 An engine block that has been separated waits to be cleaned. Note how the liners on the left and right have come out of the cylinders slightly.

5 A camshaft is replaced.

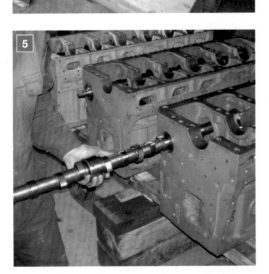

6 Carburettors are rebuilt.

7 Pistons and con rods are reassembled.

8 This is the fuel distribution box pictured before its restoration.)

9 The petrol distribution box after restoration. *(All photos on these pages Author's collection)*

10 The voltage control box after restoration. *)*

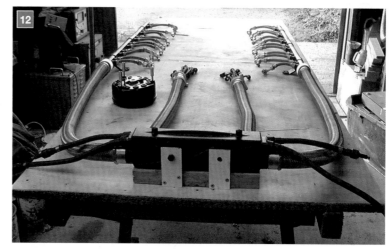

11 Part of the ignition system under restoration.

12 The ignition harness.

13 A distributor cap.

14 The instrument panel is rebuilt.

15 The battery master switch and panel, restored and re-installed.

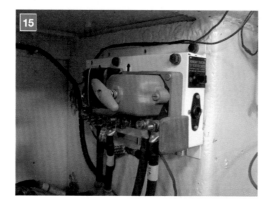

alloy. These are prone to frost and other physical damage and need to be checked, as do the two oil pumps, and the fuel pump, below the engine. Whereas the oil pumps tend to benefit from being well lubricated, the fuel pumps are liable to corrosion of the filter bowls and to jamming. The latter can be cured by a rebuild, but the bowls may need to be replaced. The fuel pump also has a Bowden cable that enables the crew to pump petrol up to the float chambers by means of a 'T'-handled pump on the bulkhead at the back of the fighting compartment (see page 113).

The gearboxes tend to survive decades of non-use, providing that they are uncracked and closed. They will need flushing and checking, and care must be taken to select gearbox oil that is compatible with the bearings and seals in these older mechanisms. The compressor on top of the gearbox needs to be dismantled and inspected, as well as overhauled.

Both sets of brakes – the steering brakes on the gearbox and the main brakes on the hull, inboard of the final drives – need to have the drums removed, the slave cylinders reworked and the shoes looked at.

The clutch will most likely need to be rebuilt; corrosion tends to damage the springs and fingers and the whole unit will require careful removal and restoration (in one restoration, the clutch was accidentally dropped and, although a substantial item, it broke). The clutch friction plate seems normally to be in good condition, although do bear in mind that, along with the brake shoes, it is likely to be made with asbestos.

The 'Sirocco' fan is likely to need restoration as well. This is a very large fan that sits by the clutch on the flywheel and draws air via the external air louvres and through the radiators

ABOVE This is the underside of a restored engine. On the left, behind the starter motor, can be seen the two bowls of the petrol pump. To the right of these on the opposite side are the two sumps.

LEFT A gearbox with the steering brake drums removed. The friction lining can be seen clearly on the one on the left. The finned box on the top, right, is the compressor.

LEFT The gearbox, right-hand side, showing the brake drum off and the brake shoes and expanders. Hanging down on the right are the selector rods.

FAR LEFT Clutch components after blasting and cleaning.

LEFT Here is the clutch fitted to the engine.

RIGHT The Sirocco fan under restoration.

RIGHT On the left is a rebuilt oil cooler; beside it is one of the two radiators for one side of the engine compartment. The opposite pair can be seen behind.

FAR RIGHT The right-side radiators and oil cooler in place inside the engine compartment. The pipe emerging from between the base of the two radiators is the petrol pipe from the fuel tanks, which sit outboard of the radiators.

and oil coolers back across the engine and gearbox and out of the rear of the tank. The fan rotates at the same rpm as the engine and needs to be structurally sound to avoid it being destroyed by centrifugal force. The bottom of the fan is often corroded from sitting in water lying in the engine compartment. Some have been damaged in other ways and most by now are corroded to a point where they are best rebuilt, rivet by rivet, and then balanced.

The two radiators and one oil cooler on each side are generally suspect. Radiators of that age that have been residing outside for decades tend to leak and even the oil coolers are not likely to be reliable (one restoration where the original oil coolers were reinstalled leaked when the engine warmed up; as a result, the engine had to be removed again, in order to replace them). It is recommended to rebuild the radiators and oil coolers, replace the pipe work and, if necessary, make new mounts and brackets for them.

Corrosion is a problem in the fuel tanks. There are three fuel tanks on each side of the tank, and condensation forms on the inside. This often leads to pin-sized holes appearing in the metal, and even if not visible to the naked eye, this should (but does not always) show up on compression testing of the tanks. In our case fuel tanks were remade with a form of stainless steel, reusing the old connections and fittings with new pipes. The rubber in the old pipes had cracked and perished; in any event, given the chemical constituents of modern fuels, it could be hazardous to use old rubber pipes. The 'sender' units for the fuel gauges will almost certainly need to be overhauled. These sit in the vertical tanks and all have proved

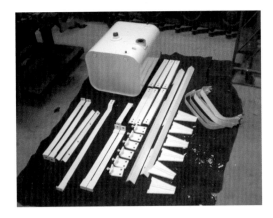

ABOVE One of the vertical fuel tanks with its associated bracketry.

RIGHT The top diagram shows the layout of the fuel system (and will make the accompanying photographs easier to follow). The lower one explains how the fuel system works.

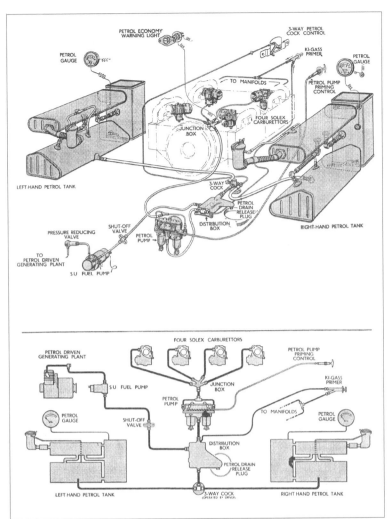

capable of working after dismantling, cleaning and rebuilding.

The hydraulic or, more correctly, air over hydraulic, system used for control of the steering brakes and clutch always requires overhaul. This consists of replacing the pipes that run from the driving position to the engine compartment. (On our project all of the various pipe unions and joints were replaced by a single run of pipes from the driving position to the engine compartment.) More importantly, the process involves removing and dismantling the 'pedal cluster'. This is the box in front of the driver, level with his knees, bolted to the glacis plate, which is a hydraulic oil reservoir tank below which the accelerator, brake and clutch hang, as well as the steering controls and all the associated master cylinders and the air pressure gauge. Rubber seals in the master cylinders tend to have perished, and quite often the pistons will have scored, or corroded in, their bores. The assembly is a sensitive one, and if not correctly restored and set up, will

RIGHT The lower fuel tank on the right-hand side of the engine compartment is linked to the distribution box on the engine compartment floor. This distribution box enables the front gunner to select right or left fuel tanks.

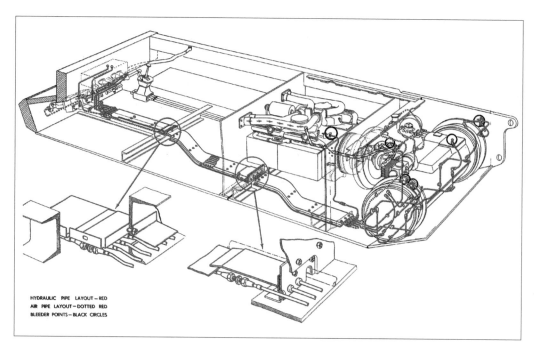

HYDRAULIC PIPE LAYOUT – RED
AIR PIPE LAYOUT – DOTTED RED
BLEEDER POINTS – BLACK CIRCLES

RIGHT **Pipework in the Mark IV gearbox compartment.**

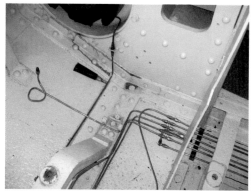

lead to faulty steering. For this reason, it is best to bench-test the rebuilt pedal cluster with compressed air, particularly to ensure that the steering will work as it should – unpredictable steering in a tank is hazardous, to say the least!

All wiring is likely to need replacement including the engine ignition harness, which is installed in screened conduits to prevent radio interference. The insulation on all of the original electrical cables is prone to perishing and given the potential fire hazard in the event

RIGHT **The wiring layout for the driver's and fighting compartments.**

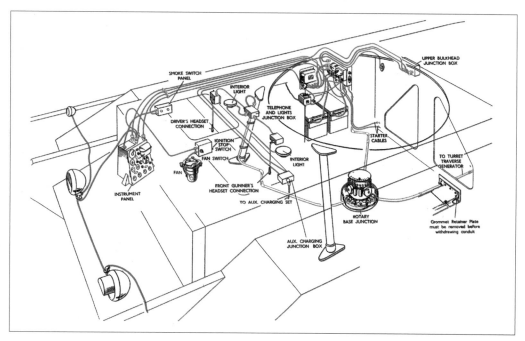

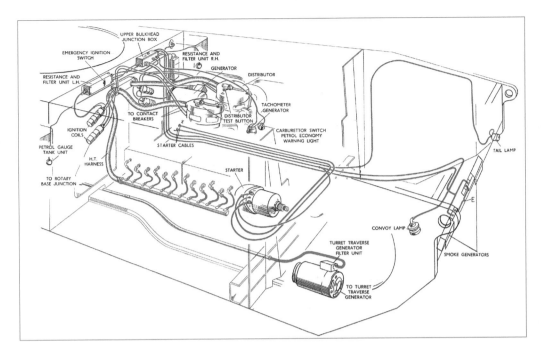

of a short circuit, it would be foolish to run on
the originals. The internal wiring is generally
protected by a flexible conduit and, in the
case of the radio cabling, a mesh screen
similar to that used on the ignition system. This
means that modern wires will not be visible
inside the tank.

Externally, the guide channels that keep the
tracks in place as they run back along the top
of the panniers tend to corrode and lift off the
hull. These are similar to large sections of angle
iron, although in fact they are not cast at a right
angle, but a slightly larger one.

Front idlers and final drives need checking
and, although normally sound, will require
lubrication. Track links are inclined to seize
over time, creating long runs of plank-straight
track, and these need to be gently worked on
to make them flexible again. Heating the track
as a short cut to achieve this seems to produce
disastrous results, as the links become fragile
and prone to breakage. Slowly towing the hull
with the track fitted appears to work. Some pins
will break in the links and require considerable
persuasion to come out.

Traverse gearboxes, which enable the
gunner to rotate the turret, have a habit of
corroding internally, but can be restored.

In a full restoration, the turret will need to be
removed to allow shot-blasting of the hull and
replacement of rusty metal in the floor areas.

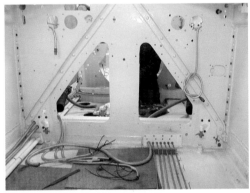

LEFT Hydraulic
and compressed
air pipes leading
through from the
engine compartment
(on the other side of
the triangular cut-
outs) to the driver's
compartment.

LEFT A replacement
rotary base junction
(this one is from a
Scorpion CVRT), which
enables power and
intercom signal to
move from the rotating
turret to the hull.

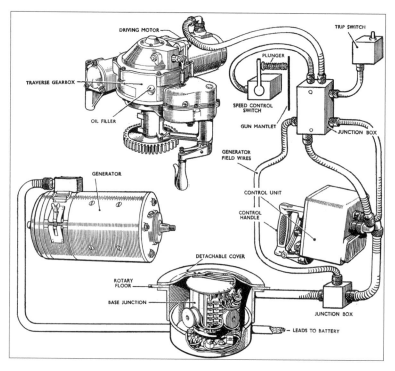

of 14 segments. In either case, the teeth may be damaged or missing, and this should be checked and, if possible, rectified. Although not necessary in our case, it would be possible to have a replacement section of ring made and machined. Another small but vital element of the traverse mechanism is the turret lock. This is a small bolt that pushes up from below the turret bearing and locks the turret to the hull. If the tank is to be driven, this bolt should take the strain of the turret trying to rotate as the tank drives over rough ground, unless the traverse gear is being used.

The main gun and breech mechanism are normally fairly robust, being made of substantial lumps of metal. The breech ring and associated assemblies can be removed (in theory, the breech ring unscrews from the chamber end of the barrel; in practice both tend to be very firmly stuck together – take care if dismantling them as the breech ring is extremely heavy) and the barrel can be removed via the hole in the rear of the turret for cleaning and restoration. There is a recuperator on top of the gun cradle – a spring damper that returns the gun to its forward position after recoil – and it is sometimes corroded inside the casing. The spring itself is not a visible component, but it is the means of preventing the gun from sliding backwards into the turret, so care should be taken to ensure that it provides the necessary security. Do not forget that the gun will, in many countries, be classified as a weapon, even if it is of an obsolete calibre and not complete. Always check with the appropriate authorities and if required, deactivate the weapon to the necessary standard or obtain permission to own it.

ABOVE The traverse gearbox and associated components. Note that the electric traverse mechanism also needed the rotary base junction in order to work. *(All illustrations on these pages Author's collection)*

Access to most tasks in restoration is easier with the turret off. The turret rests on a large circular bearing comprising an inner and outer ring separated by 117 ball bearings held by 13 strips of phosphor bronze with holes in them. These ball bearings, of 1.25in diameter, will frequently have rusted but can readily be replaced. One question with the reassembled turret bearing is whether to grease it. From experience (and as the manual instructs), if you do you will find grease dripping into the fighting compartment for a long time afterwards! The outer portion of the bearing – the part that is attached to the hull – has teeth set into the inner face, and the traverse gear grips these to enable the turret to turn. On later Mark VII and VIII tanks this ring is sometimes formed not of a single casting, but

RIGHT The Mark IV turret after shot-blasting and priming.

FAR RIGHT The Mark IV turret, pictured upside down with the gun mounted, but without the breech ring.

The front Besa machine-gun mount is likely to have corroded and be stuck in one position if the tank has lived outside. The pre-Mark VII tanks had a mechanism for providing movement in this fixture comprising one pair of bearings above and below the Besa mount, and another pair on the sides, much like a universal joint. Once cleared of corrosion and lubricated, the whole mount can swing freely within the relatively limited arc of movement permitted by the hull. In the Marks VII and VIII, the mount takes the form of a giant ball and socket, which seems less prone to corrosion but still needs to be dismantled and lubricated. These mounts are only balanced once a Besa is installed, as they are designed to allow for the weight of the gun.

The commander's cupola and the hatches on the turret, hull and engine/gearbox compartments are often corroded, along with the frames that they sit in. These are best removed, renovated where necessary and then reinstalled. The spring catches that hold the turret hatches open are external components and prone to rust. It is essential that these are restored, as one of the hazards of Churchill (and most) tank use is falling hatches and damaged fingers. On the later all-round vision cupolas, the spring assemblies that help to open and close the hatches tend to rust solid and will require restoration.

On all marks, the driver's vision port and the pannier doors will need to be freed and lubricated. The pannier doors can be immensely difficult to open, as they are tapered, and with rust and time they wedge shut. One technique for dealing with this is to use a powerful hydraulic jack braced between the doors to push them from the inside. The vision ports, which are more complex on the pre-Mark VII tanks due to the two-part 'door', often need to be dismantled to free the bearing on which the door rotates and the release mechanisms set in the armour. Both the pannier doors and the vision ports should have bare metal on the tapered closing surfaces, which ought to be lightly oiled. This helps to prevent them from jamming.

On what might be termed 'square door' hulls – that is, all marks prior to the Marks VII and VIII – the side armour of the hull is bolted on to a mild steel inner. Over many years water finds its way between the two and, as rust forms, it pushes them apart causing enormous pressure and even

LEFT **The 75mm gun under restoration.**

forcing the sides to bow perceptibly. As the Mark VII hulls were manufactured in one piece, they do not suffer the same problem. On our Mark III*, the armoured sides were removed to enable the steel inner to be rust-treated and repaired.

The track guards and external 'tin work' is unlikely to have survived decades of abandonment. In our case, track guards were remade to original specification, which is a painstaking business and an occasional subject of controversy.

On hulls and turrets that have suffered shell or other projectile damage, repairs may be needed. Our Mark III had suffered an explosion between the turret and the hull roof, just between and behind the driver's and front gunner's hatches. This had blown the roof section down, and the base of the turret up. The turret itself had received a shell impact next to the commander's cupola, which had blasted the roof down and split it. The pannier door on one side had been blown off, causing some damage to the inner hull. All of these areas were reworked and restored. On the Mark IV, there was damage to the sides of the turret caused by shell splinters and bullets. This has been left visible, but the sharp edges created where the metal was gouged were ground off to prevent injury.

BELOW **The track guards are fitted.**

Even without shell damage, the hull and turret of any tank that has been sitting outside for decades will have rusted. Our project entailed stripping out all of the fittings and sandblasting the hulls and turrets before priming them (grey primer on the inside, to go under white paint; red primer on the outside to go under the 'olive drab' paint). The first layers of paint on the outside were gloss, to give a water barrier. Some of the mild steel partitions and fixings on the floor had to be replaced, as water and moisture tend to gather here most of all.

Approaches to historical accuracy

This is an issue that must haunt all restorers, not just of Churchill tanks. A number of key issues need to be confronted:

■ What point in a tank's history will you take as your basis for restoration? Aside from its different roles during the Second World War, it probably served afterwards as well, with different equipment, colour schemes and even functions.
■ Having made that decision, is there such a thing as a 'correct' set-up for everything from the fittings to the colour and markings?
■ What compromises are you prepared to make to originality in the interests of safety, ease of use and economy?

Every restorer will respond to these questions differently, and indeed one's views on some aspects of them can change as the work progresses. However, a restoration of a

historically significant vehicle like the Churchill tank carries some duties to those who may see the vehicles and indeed own them in the future.

The Churchill Tank Project machines are being restored to correspond to different points during the Second World War. Records in the National Archives make clear that the Mark IV was a 6-pounder gun tank before being turned into a 'Twin ARK' after the war. That makes some decisions easier – it has been returned to that form, leaving only a question over whether to have geared elevation or 'free' elevation, and what cupola to have for the commander. Studying the surviving records at the Tank Museum led to a conclusion that most 6-pounders had free elevation and, also, early two-periscope cupolas for the commander. From a personal perspective, too little historical emphasis seems to be placed on the fighting in North Africa and Italy, and it has been decided to restore the tank to the standard and fittings that would have been found in those campaigns.

The Mark III had appliqué armour fitted, and had an all-round vision cupola, leading to the opinion that it would be best set up as a tank that had taken part in the fighting in north-west Europe after the D-Day landings. There were both 6-pounder and 75mm Mark III* tanks in these campaigns, but it was decided that this tank was to be a 75mm one, with geared elevation and mechanical (as opposed to electrical) firing gear.

The Mark VII did most of its fighting in the same campaigns and it too is set up as a post-D-Day tank. As all Mark VII tanks were fitted with the 75mm gun and electric solenoid firing equipment, so too is our Mark VII.

All of the tanks are painted in 'olive drab'. The Marks III* and VII will have the American star painted on the turret to facilitate recognition by aircraft as Allied tanks, as this appears to have been fairly standard, but by no means universal, practice. The interior of all the tanks is white. There is some debate about whether silver (the standard post-war interior colour for British tanks) was used during wartime. Having interviewed Churchill veterans and researched the issue, my conclusion is that (i) white paint was certainly used, although it cannot be proved that silver was not; and (ii) quite unlike the German system of prescriptive colour codes, the factories in Britain were told to use such light-coloured paint as was available. The engines are painted black and the gearboxes dark gloss green, as research suggests that these were the correct colours during the Second World War for British tanks. The use of 'eau de Nil' – a kind of light greenish blue – on engines in British tanks was a post-war practice and we have Churchill engines in this colour, have all been rebuilt since the war.

The Wireless Set No 19 is also something of a conundrum. There are three main variants: the Mark I, Mark II and Mark III. Some have Russian script as well as English markings. As far as I can establish, these were manufactured for use by Russian troops under the Lend-Lease programme, but were also used by the British. The Mark IV has the WS38 installed in front of the gunner's position, as its turret had the appropriate holes drilled for an aerial for such an installation. More details about the radios and their associated equipment is found on page 89.

Is there such a thing as 'correct' when it comes to fittings, colour and markings? The general belief is that, outside the major items (calibre of gun, engine, gearbox and the like) there is not. Tanks were worked on in the field and adhoc changes were made, unofficial additions created and stowage altered to suit the crews and the current campaign. As mentioned above, colour on the British tanks was not subject to rigorous specification, and markings moved position and size regularly. Thus, in what must always be a personal perspective, I used photographs of tanks in the correct campaign to see what fittings would be present is a good guide, and these show that the stowage lists and manuals are often misleading.

These are controversial areas, particularly for model makers, who often have very strong views about what is 'right'. The more research is done, the more convinced one becomes that there is seldom just one 'right'!

Many tanks were festooned with additional stowage – a habit perhaps first seen with the Eighth Army in the desert, but frequently seen in north-west Europe. Ammunition boxes were often fitted to the sides of the turret to provide space for the crew's belongings, and these were attached by welding threaded bosses to

the turret and then bolting a box that had once held ammunition to the sides.

Informal additional armour was often added, generally by welding track links to the turret sides and parts of the hull. Plates welded to the turret to prevent sniper fire from hitting the commander were to be seen on some tanks in Normandy.

Compromises? Any restorer will be faced with the dilemma of what to do to maintain originality in the face of cost and scarcity (or complete absence) of original parts and in a world where some of the original materials are unsafe to reuse.

My simple philosophy on this is, first, to try to keep examples of any item that is being replaced by a modern counterpart. Second, safety is always put ahead of originality where an item might be dangerous in reuse. One example of this is wiring and pipes. It would be very risky to reuse the original wiring, even where it has survived. Fortunately, almost all of the wires run in conduit, or, in the case of radio cabling, in conduit with a metal screen over the outside. Modern cables that are fire resistant have been used in our restoration work and these cannot be seen inside the conduits, which themselves are new. Hydraulic, petrol and other pipes are not safe to reuse and have been replaced, but originals have been kept for future reference. The rotary base junction can be a relatively modern one, of the type used in CVRT Scorpion light tanks. An original is in our stores to act as a reference.

Many items can be found by diligent and patient research and requests. However, not all will turn up in time. Much of what might be termed 'tin work' – stowage bins and fittings inside the tank – has rotted away, and in our restoration, where replacements could not be found in militaria shows and on old tanks, replicas were fabricated using what was left of the originals as patterns. The originals are retained, even though badly rusted or damaged, to act as future references. Occasionally a guess is necessary as to how a part was made, as there are few original engineering drawings left. This has only been done where an original example cannot be traced, and in such cases photographs, drawings in manuals and comparable tanks of other types that survive in museums have been used as references.

Chapter Six

Servicing and maintenance

The Churchill is not an easy tank to run in modern times. It is one thing servicing and maintaining it with a crew of five and a REME team in support, but quite another doing it without all that back-up. These are some of the key tasks and the ways in which they can be done.

OPPOSITE A well camouflaged Churchill Mark IV with a 6-pounder gun, possibly of 51 RTR, is given a final mechanical check over in a harbour area of Italy, 20 July 1944. Note that the track guards are missing on the centre section and at the front. The crew member on the left of the photograph is sitting on 6-pounder ammunition boxes, with a wooden Besa ammunition box in front; this wooden box would have contained two metal liners, each with a full belt of ammunition. The driver can be seen adjusting the track tension while the turret crew pass an oil can between them, quite possibly to lubricate the hatches. There appears to be three aerials on the back of the turret, the middle one being for the 'B' Set part of the WS19. This means that the tank probably had a WS38 fitted as well. The tank has been well camouflaged. Note the height of the trees compared to the hull Besa, particularly in the light of the comment about 'close-country fighting' on page 147 and the weapon being low enough to use in wooded areas. *(IWM TR2017)*

General description

In military service, maintenance was principally the function of the driver, who had attended a lengthy 'Driving and Maintenance' or 'D&M' course. In fact all of the crew assisted and although they had their own assigned tasks, each was trained to be able to do the jobs of the others in case one was injured.

The tanks carried a toolkit just behind the driver's seat. With these, normal maintenance would be possible. More demanding tasks would be carried out by fitters from the Royal Electrical and Mechanical Engineers (REME) who generally had Churchill Armoured Recovery Vehicles (ARVs, see Chapter 2) to carry their equipment, together with welding and lifting gear.

BELOW **Lubrication chart – Churchill Mark I, II, III and IV.** *(Author's collection)*

OPPOSITE **Lubrication tasks explained.**

Regular maintenance operations

In addition to the checks noted in Chapter 7, the following lubrication tasks should be carried out. The original oils may no longer be available and modern replacements will then need to be considered. This must be done with care, as a number of modern oils and greases are extremely detrimental to older bearings, metals and seals. Always seek guidance from experts before using any oil or grease.

Some of the tasks listed may appear to be unnecessary on a tank that is rarely used (the original service intervals are noted with the tasks), but other than those in relation to gun parts, they should be borne in mind at least at

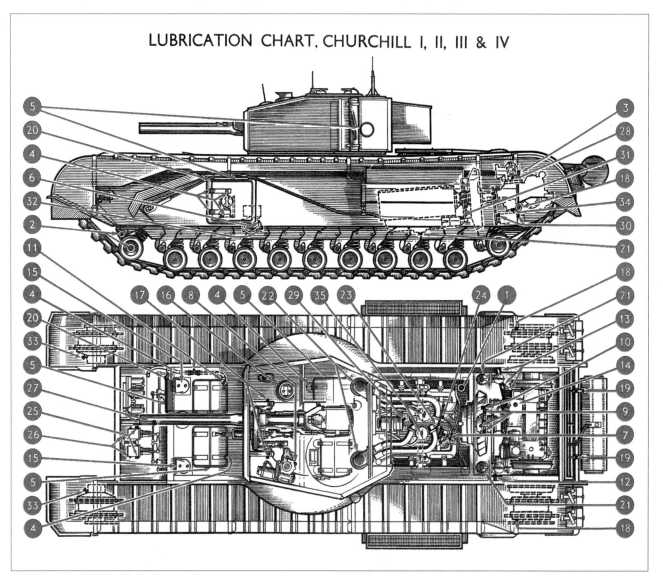

LUBRICATION CHART, CHURCHILL I, II, III & IV

ENGINE OIL FILLER AND DIPSTICK

BOGIE AXLES AND FULCRUM SHAFTS

AIR COMPRESSOR

VISION PORT

AUXILIARY GENERATOR

THROTTLE CONTROL ROD JOINTS

TURRET TRAVERSE GEARBOX

CLUTCH THROWOUT MECHANISM

SERVO MOTOR CLEVIS JOINTS

HYDRAULIC RESERVOIR FILLER AND LEVEL PLUGS

SPEEDOMETER DRIVE BRACKET BEARINGS

CHANGE-SPEED CONTROL ROD JOINTS AND LEVERS

GEARBOX FILLER PLUG AND DIPSTICK

PERISCOPE

GUN TRUNNION BEARING

FINAL DRIVE FILLER PLUG

AUXILIARY TANK RELEASE MECHANISM

LOADING DOOR NIPPLE

ENGINE DRAIN PLUG

AIR FILTER

GOVERNOR

TACHOMETER

STEERING SERVO CYLINDER

HAND BRAKE MECHANISM

AIR COMPRESSOR

MAIN GENERATOR

STARTER

AUXILIARY GENERATOR

IDLER WHEEL

GEARBOX DRAIN PLUG

DISTRIBUTOR AUTOMATIC MECHANISM

DAILY
■ Air compressor
Check oil level in crankcase and top up – fill to top of filler hole.
■ Auxiliary generator
Check oil level if used. Top up to lower thread of filler hole.
■ Vision port and pannier doors
Lubricate the opening and locking mechanism of vision port with engine oil and lubricate round edges of vision port and pannier doors with gear oil.
■ Hatches
Lubricate hinges and locking catches with engine oil.
■ Bogies
If 75 miles or more have been covered since last lubrication, treat with gear oil. In our case, grease has been easier to manage for this, given the tendency of the seals to weep and the limited mileage travelled in the tanks.

WEEKLY
■ Engine
Lubricate all throttle control ball joints and linkages with engine oil.
■ Clutch
Throw-out mechanism – three lubrication nipples for gear oil.
Clevis joints – oil these joints with engine oil.
■ Gearbox change speed rods
Lubricate each of the gate mechanism and rods in the gear-change lever bracket by the driver, as well as the control joints, actuating levers and sliding rods leading to, and in, the gearbox compartment, with engine oil.
■ Hull periscopes and front Besa mounting
Lubricate the two nipples on each periscope with gear oil.
Lubricate the clip and hinge pin on each periscope with engine oil.
Lubricate the front gun-mounting nipple with gear oil.
■ Seats
Lubricate all seat hinges, catches and slides with engine oil.
■ Hydraulic system
Check fluid level by looking at the level plug in the filler pipe. This plug is roughly halfway down the pipe between the cap in the driver's hatch and the reservoir in the pedal cluster.

■ Gun mounting, depression and firing gear
Lubrication details vary; essentially use engine oil on moving parts and gear oil on the two nipples for the actuating shaft and the nipple for the spring case.
■ Gun travelling lock
Lubricate at each end of the lock with engine oil.
■ Turret race, inner ring
Lubricate the six nipples on the inner race (later tanks) with gear oil while slowly traversing turret, till oil oozes out below the race all the way round.
■ Turret lock
Lubricate plunger with engine oil.
■ Turret periscopes
As for hull periscopes (see above).
■ Turret traverse gear
Check oil level in the gearbox and top up as necessary using engine oil. Lubricate the nipples on the gearbox casing with gear oil. Use engine oil on joints, spindle, pivots and visible moving parts.
■ Engine and gearbox compartment hatches
Lubricate hinges and locking catches with engine oil.
■ Gearbox
Check oil level and top up if required with gear oil. The dipstick to measure the level is next to the input bevel housing. Screw it fully in to inspect level and top up using either of the filler plugs. Check that both air vents are free from obstruction.

MONTHLY
■ Idler units
There are two idler wheels, each with one lubrication nipple. Lubricate with gear oil until oil oozes out at each side of the idler hub.
■ Steering mechanism
Lubricate the servo cylinder nipples with gear oil and the two oil cups with engine oil.
■ Auxiliary generator
Drain sump and refill with engine oil.
■ Handbrake mechanism
Lubricate pivots and ratchet release with engine oil.

EVERY 500 MILES
■ Distributors
Oil automatic mechanism through the holes in each distributor arm with engine oil and wipe the arm dry afterwards.

MONTHLY OR EVERY 500 MILES, WHICHEVER COMES FIRST

■ Final drive units
Lubricate the two final drive assemblies through the filler plugs (accessible via a hole behind a removable plate in the side of the hull). Remove the plate and turn the final drive unit until the filler plug is opposite the hole. The oil level should be at the lower edge of the filler plug hole. Use gear oil.

■ Governor
There are two grease cups to lubricate the bearings for the external control rods. Give each cup one complete turn. Replenish grease as necessary.

■ Tachometer generator
Screw down the grease cup by one turn. Replenish grease as necessary.

■ Starter motor
Remove and fill the oil cup with engine oil.

EVERY 2,000 MILES

■ Gearbox
Drain gearbox and refill to level mark on dipstick with gear oil.

To drain, first remove the drain plug from the sump at the lowest point on the gearbox. The flow of oil will flush the internal filter gauze and should remove sediment from the sump. Once this flow ceases, take out the other two plugs from the bottom of the gearbox to make sure all oil has come out. These two plugs are magnetic to attract and hold any metal particles. Each should be put back only in the hole from which it came, and only after careful cleaning. When refilling the gearbox, take out both filler plugs as this will speed up the process. However, allow time to elapse before checking the level, as the oil takes a while to find its level. Make sure that you screw the dipstick firmly back after use.

regular intervals, particularly when the tank has been running, or may have been damp from rain or condensation.

Even tanks stored indoors are prone to attracting condensation both internally and externally. A dehumidifier placed near the tanks to control this when the weather is damp is a good idea, as well as keeping all hatches open to avoid build-up of moisture on inside surfaces.

Clean water drains on top hatches
Open the driver's, front gunner's and loader/operator's hatches and secure them. Around each hatch is a drainage channel; on the front hatches there are four drain slots. On the loader/operator's hatch there is a single hole at the right rear corner. Use a piece of wire to clean these. The loader/operator's hatch drain hole has a rubber pipe leading to a metal one. This too may need clearing.

The commander's cupola has a drain channel and drain slots that should be cleared.

BELOW Cleaning the hatches. (Author's collection)

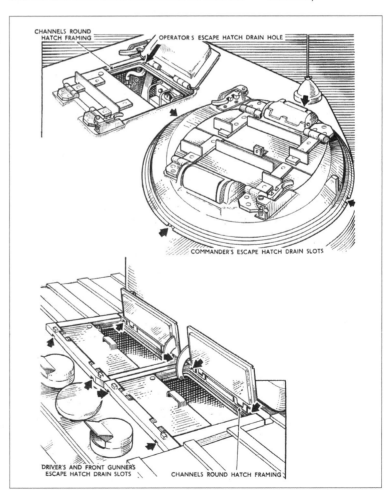

CHANNELS ROUND HATCH FRAMING

OPERATOR'S ESCAPE HATCH DRAIN HOLE

COMMANDER'S ESCAPE HATCH DRAIN SLOTS

DRIVER'S AND FRONT GUNNER'S ESCAPE HATCH DRAIN SLOTS

CHANNELS ROUND HATCH FRAMING

Check controls for tightness

Check all movable controls.

Inspect for leaks

Carrying out a periodic examination of pipelines and unions is very important; this should be done slowly and methodically, with proper lighting. Ideally, do this inspection before any cleaning, as leaks may be detected by spills or marks that would be lost when cleaning is implemented.

If evidence of a leak at a pipe joint is found, tighten the relevant union if it is loose. If it is tight, dismantle (draining the system if necessary) and check the threads and consider whether to use a jointing compound or other means of improving the situation. In some cases, it may be necessary to replace a fitting if the thread is loose or worn.

■ Petrol system

Originally all of the flexible pipes in the petrol system had wired unions that had been locked on assembly. There are three main tank interconnecting pipes that can only be checked after the air louvres have been removed from the tank (which has to be done before transporting the tank by road unless it is to be escorted as a wide load, under current UK regulations – check before any road transport is organised for current applicable rules).

Check the filler pipes visible inside the gearbox compartment.

Examine the three-way fuel tank selector valve and pipes to the petrol distribution box and from it to the petrol pump, all viewed through the inspection ports in the floor of the engine compartment. There is then a pipe from the distribution box to the petrol junction on the top of the engine, clipped to the engine compartment rear bulkhead. Four pipes go from this to the carburettors.

Check the pipe from the petrol distribution box to the suction side of the Ki-gass primer and two pipes from the primer to engine inlet manifolds.

■ Oil system

Check pipes.

■ Water cooling system

Check pipes.

■ Air control system

Check pipes.

■ Hydraulic systems

The connecting pipes from the front controls to the rear gearbox compartment are grouped in conduits and were originally made in separate sections. On our project these have been replaced with single pipe runs to reduce the possibility of leaks in inaccessible places. There are six pipes in the set and they are for throttle, parking brakes, clutch, steering left hand, steering right hand and compressed air line. Check these at each end.

■ Gearbox

Examine for leaks from joints, as well as at the input shaft and the output shafts. Check for oil thrown from the brake drums, and hydraulic fluid leaks from the pipes and operating cylinders on the steering brakes.

■ Final drives

Look for leaks from each side of the hub and from the casing joints around the outside.

■ Idler units

Inspect around the hub plates, but note that oil coming from the shaft is normal – it is the measure when lubricating.

Checking and adjusting the clutch pedal

A simple method of checking clutch pedal clearance is to run the engine until the air system is fully charged, and then briskly push down the pedal while watching the air pressure gauge. When the clutch starts to disengage, the needle on the gauge will flick. The amount of 'free' pedal travel is the movement of the pedal from the 'off' position to the point at which the needle flicks. Take care while doing this test not to overshoot the mark with the pedal. The free travel should be more than one inch and less than two.

Adjustment to lengthen the travel is carried out in the gearbox compartment, on the right-hand side, by shortening the hydraulic cylinder pushrod under the air pressure cylinder. Slacken the lock nut and then adjust the rod by turning the other nut clockwise (as seen looking to the front of the tank).

The clearance that should be maintained is the distance between the clutch fork trunnion and the throw-out sleeve, and it must lie between a maximum of $1/8$in and a minimum of $1/16$in.

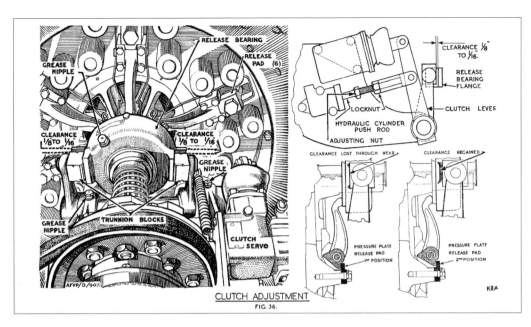

Adjusting the tracks

The tracks run over an idler sprocket, sometimes referred to as a front idler, and this is provided with an adjustment to take up track wear. Each idler is mounted on an axle carried in slides; at each side there is a nut through which the adjusting bolt operates. Between the rear of each slide and the sliding block on the axle end are packing blocks, which take the thrust of the drive. These blocks slide on an adjustment screw and are secured by a final retainer that is bolted in place. A large cap nut at each end of the axle provides a final lock. These cap nuts are held against loosening by two set-screws.

To adjust, hinge up the front section (or 'hood') of the track guard, if present. Remove the set-screws securing the idler axle nut lock plates. Slacken the axle nuts; tighten the adjusting screws evenly and remove the set-screws holding the retainers. This enables the packing blocks to be extracted. Further tightening of the adjusting screws (equally on each side) will bring the idler axle forwards and take up slack in the track. The adjustment is judged to be correct when, with the slack portion of track at the front, it is just possible to rotate the bogie wheels of No 1 suspension unit by hand.

When the required adjustment has been achieved, fix the requisite number of packing blocks to fill the space, until only the retainer can just fit in. Secure the retainer and slacken back the adjustment screws to relieve tension on the threads. Finally, retighten the axle cap nuts and secure with the locking plates. Return the front section of the track guard to its proper position and secure with its bolts.

Cleaning engine air and fuel filters
■ Air cleaners

There are two air cleaners fitted inside the rear of the fighting compartment. These provide air for each of the cylinder banks, although the left-hand unit also supplies filtered air for the

Fig. 82. Exploded view of AC two-stage air cleaner.

(1) Top casting. (2) Gasket. (3) Wing nut. (4) Filter elements.
(5) Compensator chamber. (6) Oil pan. (7) Central vane assembly.
(8) Sealing ring. (9) Air cleaner casing. (10) Bottom dust container.

compressor. Early tanks had a single-stage filter and this was replaced with a two-stage version. On the Churchill Tank Project the internal components of the filters have been replaced with modern air filters. What follows is the routine for those who retain the original system, although plainly the modern variety need to be checked and maintained as well.

Slacken the lower wing nut holding the canvas bag to the cleaner. The container can then be removed, a manoeuvre that is performed by traversing the turret until a cut-out section of the platform ring shield is next to the filter – this allows it to be removed without tilting.

To clean the oil-bath chamber and filter elements, traverse the turret until the commander's seat is (for the right-hand cleaner) towards the rear of the tank and (for the left-hand cleaner) at the left side of the tank. This allows the cleaners to be removed with minimal tilting. Remove the dust container as above and empty it. Unscrew the side wing nuts and lower the cleaner body until it clears the casting that holds it. Withdraw the whole body, but avoid tilting as much as possible, to prevent oil spillage.

Lift out the oil bath, keeping it vertical. Unscrew the central wing nut and lift off the element assembly. This element is made of metal-wool and should be cleaned (originally in petrol before dipping in engine oil). Dismantle and clean the remaining components,

check the seal and gasket and when ready, reassemble.

■ Petrol filters

There are two gauze strainers in the petrol fillers, which are in the gearbox compartment. These should be cleaned thoroughly.

There is a strainer for the supply to the petrol pump inside the petrol distribution box on the engine compartment floor. Remove the right-hand front access plate to get to this. What follows is the original procedure for cleaning; it may now be considered too hazardous to carry out: 'Turn the petrol supply to "off". Remove the drain plug from the distribution box and flush the box out by having someone turn the petrol supply to RH. or LH.'

Lastly, there are two gauze strainers under the petrol pump, enclosed in a metal bowl. Turn the petrol supply to 'off'. Remove the large rectangular access plate in the engine compartment floor. Loosen the stirrup holding each bowl and remove the bowl but not the strainer. Flush out the bowls and replace, making sure that they form a tight seal. When retightening the stirrup securing nuts, do so only by finger pressure, to avoid damaging the gaskets or bowls.

Check bolts on bogie assemblies

Inspect the bogie bracket tie-plate bolts regularly.

Adjust fan belts

There are driving belts for the engine dynamo, water pumps, the traverse dynamo and the speedometer drive.

For the engine dynamo, mounted on top of the engine at the front, the adjustment is made on the left side of the dynamo. To tighten the belt, slacken off the top nut and adjust the lower nut.

There are two water pumps on the front of the engine, each with its own belt. These are accessed from the fighting compartment, through the bolted panels in the rear bulkhead. Each belt has an adjustable idler pulley. To tighten the belt, slacken the clamping bolt under the pulley, loosen the locking nut above the bracket and then tighten the lower nut to move the pulley upwards. The correct tension is reached when you can depress the centre

BELOW **The fan belts at the front of the engine, which are accessed by removing the two triangular plates from the bulkhead at the back of the fighting compartment.** *(Author's collection)*

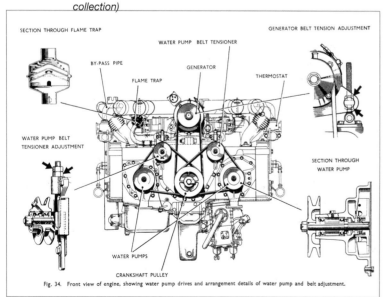

SECTION THROUGH FLAME TRAP

GENERATOR BELT TENSION ADJUSTMENT

WATER PUMP BELT TENSIONER

BY-PASS PIPE

GENERATOR

FLAME TRAP

THERMOSTAT

WATER PUMP BELT TENSIONER ADJUSTMENT

SECTION THROUGH WATER PUMP

WATER PUMPS

CRANKSHAFT PULLEY

Fig. 34. Front view of engine, showing water pump drives and arrangement details of water pump and belt adjustment.

of the belt about ¾in towards the pulleys. Lock the pulleys when finished.

The traverse dynamo is mounted on the floor of the gearbox compartment and is driven by a belt running on a pulley on the clutch coupling. Adjustment is made by a threaded rod, the head of which is at the level of the gearbox hatch. Turning the hexagonal head clockwise tightens the belt. When the correct tension is reached, it should be possible to depress the upper side of the belt by about ½in.

The speedometer in the instrument panel is mechanically driven by a cable inside an outer sheath. This in turn is driven from a pulley mounted on the left-hand final drive coupling, accessed via the left-hand gearbox compartment hatch. Tightening the belt is done by swinging the cable pulley hanger on its fixing bolts. When the correct tension is reached, it should be possible to depress the upper side of the belt by about ¾in.

Check the batteries

These are situated in the battery compartment on the right-hand side of the fighting compartment. They are likely to be modern batteries, and probably 12V ones connected in parallel. Manufacturers' instructions for battery maintenance should be followed, and the terminals should be checked for tightness and treated to prevent corrosion build-up.

Inspect the fire-extinguisher system

The procedure for this will vary according to the system you have fitted, and manufacturers' instructions should be followed.

Examine the track skid rails

The track skid rails not only serve to keep the track aligned as it passes along the top surface of the panniers, but also to prevent wear to them. To check, lift the front and rear sections of the track guards (if fitted). If necessary, slacken the track.

The greatest wear in the skid rails generally occurs at the ends, front and rear. Use a crowbar to lift the track at the ends and inspect the thickness of the skid rails. These must be renewed if the bottom flange has worn down to about ⅛th of an inch.

Check the operation of the turret traverse gear

Confirm that the traverse gearbox is securely tightened and that the teeth are meshing correctly. If you have retained the electric elements of the traverse system, examine these and the operation of the depression stop and slow speed switches. These are intended to prevent the gun barrel from hitting parts of the tank if the turret is traversed while the gun is depressed.

Inspect and adjust the steering brakes

Adjustment of the steering brakes is generally needed because difficulty is experienced in changing gear. This happens because the clutch pedal operates a clutch stop mechanism interconnected with the steering brakes. As these brakes wear, so the clutch stop operating gear becomes less effective and eventually ceases to work. The life of the clutch stop was designed to be half that of the steering brake adjustment, so problems in changing gear will be noticed a long time before the need to adjust the steering brakes becomes critical.

The steering bar has a small amount of free travel in each direction and this must be maintained for the sake of safety and operating efficiency.

Make sure that the engine is not running, and that the battery master switch is 'off'.

To check and adjust the steering brakes, first remove the two inspection plugs in each of the drums. Rotate the drums until the inspection holes come into line with the brake shoe adjusters. This rotation can be done by hand as long as an assistant sits in the driver's seat and depresses the clutch pedal sufficiently to free the clutch.

Turn the adjusters in the direction of the arrow on the drum, until the shoes press hard against the drum. The assistant should then work the steering bar back and forward at least three times to make sure that the shoes are centralised. The adjusters may then be retightened. Once this has been done, slacken back each adjuster by four or five notches.

Replace the inspection plugs and their gaskets.

Checking and adjustment of main brakes

Any need for adjustment of the main brakes will be apparent from the travel of the brake pedal, and should be made before the travel reaches its maximum.

With the engine off, remove the two inspection plugs on each drum. An assistant should then drive the tank using reverse or first gear, until the inspection holes line up with the adjusters. *This is a dangerous process and should only be carried out by competent personnel who are familiar with the risks and procedure. Clear communication must be available at all times with the person driving.*

Switch off the engine and turn off the battery master switch.

Turn the adjusters in the direction of the arrows on the drums until the shoes are tight against the drums. Have your assistant apply the brake pedal hard, at least three times, to centre the shoes, and then retighten the adjusters. Once this has been done, slacken back the adjusters by six notches.

Replace the inspection plugs and their gaskets.

Draining sumps, and cleaning and changing oil filters

■ Draining sumps

There are two sumps under the engine. Remove the drain plug from the main sump and drain the oil. It will be found that the oil tends to project forward of the sump and arrangements need to be made to catch it. Remove the centre sump

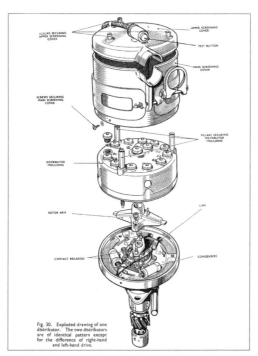

Fig. 30. Exploded drawing of one distributor. The two distributors are of identical pattern except for the difference of right-hand and left-hand drive.

drain plug as well, although not much oil should come from this one.

About 7 gallons of engine oil will be required to refill to the correct level. Run the engine for some minutes and recheck the oil level.

■ Cleaning the de-aerator filter

The de-aerator is a cylindrical tank mounted on the rear right-hand corner of the engine. It has a removable cover and contains a perforated sheet filter. Lift out the filter and clean it before replacing it and the cover.

■ Changing the oil filter elements

There are two oil filters mounted on the gearbox compartment front bulkhead, one on each side. Each contains three filter cartridges.

Remove the gearbox compartment cover plate and drain the oil from the filters. Disconnect the oil pipes and withdraw the filter bodies from the bulkhead. Take off the top covers and lift out the filter elements.

Remove the filter outlet union at the bottom of the casings and clean out the interior thoroughly.

Check the action of the bypass valve located at the top of one of the three element locating rods, making sure that it is free and returns firmly to its seat.

When finished, half-fill the casing with clean engine oil to flush and make sure that the passageways are clear.

Insert new filter elements and refit the assemblies. Do not forget to run the engine and check the oil level, as the procedure will have reduced the oil measure.

Adjustment of contact breaker points

Great care should be taken with the distributor caps and rotor arms which are fragile.

Slacken and slide back the clamp securing the high-tension wire screening on the distributor cover. Take away the three nuts securing the cover and remove it. Lift off the clips holding the low-tension leads to the hatch frame and pull them to one side to improve access.

Remove the distributor cap with the leads attached. Turn the engine (this can be done with a bar on the starter ring, if care is used and the gearbox is in neutral, and ignition off) until the rotor arm is in a position where there is clear access to the gaps and adjustment.

Ensure that the points are fully open and

slacken the screws on the adjustable arms and set to 0.010–0.102in.

Check wiring and connections for cleanliness and reassemble.

Clean carburettors

The four carburettors have the same choke and jet settings. However, their set-up does vary according to which position they are in.

Remove the air inlet pipes and front rubber hoses.

The main, slow-running and pump jets are in the sides of the carburettors, pointing to the centre of the engine. They should be taken off and blown through to make sure they are clean and free from obstruction. The float chamber should be flushed out. This can be done (with great care) by turning 'on' one of the fuel tanks and using the hand-priming control.

Disconnect the banjo connection to each float chamber, remove the gauze strainers and clean the strainers. On reassembly, check the fibre washers either side of the banjo connection and replace if necessary.

Clean radiators

It is normal for some leaves and other material to find their way through the air louvres on to the radiators. A build-up of this will make the cooling action much less effective.

Take off both air louvres. Remove any larger obstructions by hand and with a vacuum cleaner.

Blank off the louvre opening on one side of the hull using a suitable board with holes drilled to correspond to the air louvre bolts. Check that all engine and gearbox hatches are closed and run the engine at approximately 1,500rpm. With the engine running, use a brush to scrub the surfaces of the radiators on the exposed side; with one louvre opening blanked off, the flow of air will normally force smaller objects through the cores. Repeat for the other side.

If the radiators are very clogged, particularly with mud, use a hosepipe and carefully spray the obstructed areas. Do not use too much water as this can cause problems with the ignition wiring. If water has been used, open the petrol dump valve and the two drain holes. Once this has been done, and the air louvres refitted, the engine should be run at about 1,500rpm to extract any remaining moisture.

" *Remove the dead leaves and other ' foreign bodies.'* "

Removal, cleaning and adjustment of spark plugs

This is a fiddly operation that involves reaching down the sides of the engine. For this reason, it cannot really be done when the engine and ancillaries are hot.

To remove the plugs, first unscrew the combined terminal and suppressors, which should each be numbered. While doing this, hold the plug stationary with a ½in Whitworth spanner to avoid twisting the lead.

Loosen the plug using the correct plug spanner and then take out the plug by hand.

Each spark plug can be dismantled by holding the body by the hexagon and unscrewing the top, tubular, section using a ⅜in Whitworth spanner.

Clean out any carbon build-up inside the body of the spark plug and from the central electrode. Take care not to damage the porcelain. When reassembling, make sure the copper gasket fits perfectly between the two parts of the spark plug, as this joint needs to be gas-tight. Reset the plug points gap to 0.020in. When refitting the spark plugs to the engine, there should be one gasket washer in place. Our experience in rebuilding engines is that some plugs have two, almost certainly because the person doing the maintenance thought he had lost one when in fact it had stayed in the cylinder head.

Bleed hydraulic control systems

This is only necessary if air has got into the system. Generally this will result in a 'spongy' feel to the controls.

If air has seeped in, there is usually a problem either in a procedure used, or a leaking joint. When filling or topping up the hydraulic oil, take care to avoid bubbles of air being forced down with the fluid. Any leaking joint should be traced before bleeding and refilling.

Remove the filler cap and fill the main reservoir to the top. When doing this, open the air bleeder at the top left of the reservoir to allow trapped air to escape; once fluid escapes, close the bleeder and fill to the top of the filler pipe.

Do not reuse hydraulic fluid that has been drawn from the system.

For the throttle controls, locate the bleeder nipple on the throttle control chamber on top of the engine. Place one end of a rubber tube on to the nipple and the other end in the glass jar used for the purpose, having first put some fluid into the jar so that the end of the pipe will be submerged.

Unscrew the bleeder nipple about two turns and get an assistant to press the throttle pedal briskly several times, each time allowing it to return fully to the 'off' position. Continue this pumping, topping up the reservoir as necessary, until there is no more air appearing. When finished, tighten the bleed nipple, while keeping the other end of the hose below the fluid in the jar.

For the steering hydraulic system, fill the main reservoir as before. The lower and upper brake operating cylinders are interconnected by an external pipe and it is only necessary to bleed at one point for each brake. There are two bleed nipples, located on top of each steering brake support plate.

The procedure for this is the same as for the throttle, save that the pumping is achieved by moving the steering bar, and returning it to the neutral position after each pump.

For the main brake and clutch hydraulic systems, there are two main brake bleeder nipples, one for each brake, situated in the top rear corners of the gearbox compartment. The bleeder nipple for the clutch operating system is on the front of the clutch hydraulic cylinder.

The procedure for this is the same as for the throttle, save that the pumping is achieved by pressing the brake or clutch pedal as appropriate.

Removing the air inlet louvres

These louvres provide protection for the radiators, but they also extend the width of the tank considerably. Thus there are two reasons to remove them: either to carry out maintenance on the radiators or petrol tank connecting pipes or gauge sensors, or to prepare the tank for road transport. It is sensible, if doing the latter, to perform the checks set out above while the louvres are off. When restoring a Churchill, there is an option of making replacement louvres out of relatively thin steel, as the originals are extremely heavy and lifting gear is needed, along with a number of people to assist. This is a dangerous procedure, as the air inlet louvre is very cumbersome.

On original louvres, the procedure is as follows: using the lifting eyes on either side of the louvre, place a suitable chain or strop on to the lifting device and prepare to take the strain of the unit. Extract the lower bolts first and, if necessary, use a crowbar to prise the unit away from the hull.

Removal and replacement of track links or pins

Drive the tank so that most of the slack in the tracks is at the front and the link or pin to be changed is sitting between the idler and the first suspension unit. If a link is being taken off, slacken the idler and remove all of the packing pieces; let the idler back as far as it will go. If only a pin is being replaced, less slackening will be needed and the packing pieces need not be displaced.

BELOW The air louvre removed (stowed on the track guard). This gives access to the radiators for cleaning, and to the sender unit for the fuel gauge (the small silver housing on the right of the opening). *(Author's collection)*

A track puller will be required for this operation. Locate the claws of the puller in the holes on the wheel side of the track. If removing a complete link, the puller will need to bridge two pins.

The track pins are held in place by a steel strip welded to the side of each track link, closing off approximately half of the hole in which the pin is located. Normally at least two pins lack these plates, and these are the 'service' pins.

To extract the pins secured by welded strips, cut off one strip with a chisel. A tommy bar can then be used against the strip on the other side to force the pin out slightly. After this, a track pin punch will be needed to complete removal. Replacement is the reverse of this procedure, followed by welding back on a retaining strip.

Checking electrical units

These are:
- Main dynamo.
- Turret traverse motor.
- Turret dynamo.
- Starter motor.
- Rotary base junction.

The brushes of the main and turret traverse dynamos and the turret traverse motor can be examined with the units in situ. However, the turret dynamo and the starter motor have to be removed.

Take off the cover bands from the commutator end and examine the brushes and commutator. The brushes should make firm, even, contact with the commutators. Hold back the brush spring or lever and make sure that the brush is sliding freely in its holder. Any brush removed for cleaning should be replaced in the same position as it was before, as it will otherwise not make correct contact with the commutator.

The rotary base junction is located under a cover plate in the centre of the turret floor. All wiring terminals should be checked for tightness and corrosion. When replacing the cover, take great care to ensure that the driving peg is engaged in the rotary base junction and able to rotate it.

Setting ignition timing

The setting of the ignition timing is conventional

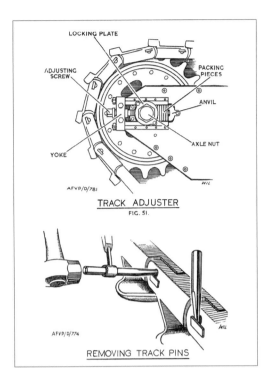

LEFT Adjusting the tracks. (*Author's collection*)

for engines of this type and era. However, each distributor rotor arm has two electrodes and there are two sets of contact breaker points in each distributor.

The following points should be watched:

With No 1 cylinder (that is, the one at the left-hand front of the engine) on the compression stroke, turn the engine slowly until the flywheel marking that indicates 3 degrees before top dead centre (TDC) on No 1 cylinder is in line with the timing pointer. (The flywheel has a line with 'No. 1 LH – TDC' and then there are three other lines before this denoting 3, 6 and 9 degrees before TDC. It is the mark nearest to the 'No. 1 LH – TDC' that must be used.)

Remove the covers from the distributor and turn the rotor arm until the end marked 'IL' is opposite the corresponding mark stamped on a boss in the base of the distributor body.

Connect a timing light to the pair of contact breaker points that serve the left-hand cylinders. These are next to the 'CL' on the base plate. With the distributor body clamping bolt slack, turn the body gently until the timing light goes out, showing that the points have opened. The timing is now as correct as a static procedure can achieve. Modern timing lights can, of course, also be used to achieve dynamic timing adjustment, but great care and skill are needed when working above the engine as it is running.

Chapter Seven

Operating the Churchill

The difference between life and death wasn't just about the thickness of armour or the power of a gun – it was also down to the skill and tactics of the crew. Read here about how the Churchill should be driven and fought, as well as what the everyday life of the crew was like.

OPPOSITE Churchill tanks of 7th Troop, 'B' Squadron, 107th Regiment Royal Armoured Corps, 34th Tank Brigade, 17 July 1944. The lead tank is *Briton*, commanded by Lt Fothergill. (See pages 138–143 for more details about this tank and its crew.) *(IWM B7636)*

In this chapter are set out two aspects of the Churchill tank; the first is the very practical side – how to start up and drive. The second part addresses the question that any restorer and historian, and indeed gamer, will ask: 'What was it like to use the tanks in war?'

How to start the tank

Unlike modern cars, starting the Churchill is something of a ritual, made easier if you have a crew of willing helpers. Even when the tanks were new, let alone when dealing with them after so many decades, there is quite a lot of checking to do before starting and driving off.

How to drive the tank

Make sure that the handbrake is fully off. Do not use first gear unless you are starting on a steep slope. Start in second gear, or if the tank is on a slight downwards slope, third gear. Engage the clutch slowly and move off slowly and steadily.

Although a small number of early tanks had a five-speed gearbox, it is very unlikely that these will be encountered, as the four-speed box appears to have been used to replace these in surviving examples.

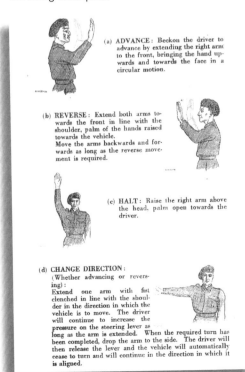

(a) ADVANCE: Beckon the driver to advance by extending the right arm to the front, bringing the hand upwards and towards the face in a circular motion.

(b) REVERSE: Extend both arms towards the front in line with the shoulder, palm of the hands raised towards the vehicle.
Move the arms backwards and forwards as long as the reverse movement is required.

(c) HALT: Raise the right arm above the head, palm open towards the driver.

(d) CHANGE DIRECTION:
(Whether advancing or reversing):
Extend one arm with fist clenched in line with the shoulder in the direction in which the vehicle is to move. The driver will continue to increase the pressure on the steering lever as long as the arm is extended. When the required turn has been completed, drop the arm to the side. The driver will then release the lever and the vehicle will automatically cease to turn and will continue in the direction in which it is aligned.

RIGHT How to guide a tank from outside.
(Author's collection)

The gear change is a 'crash' type and uses a clutch stop. With practice, quick changes up through the gears can be made using the clutch stop. This is important, as the tank loses momentum rapidly because of its high rolling resistance.

Make sure that the engine revs are up to about 1,500rpm before changing. Don't steer while making a gear change or immediately afterwards – wait until engine revs return to normal.

First gear should only be used for obstacle crossing, steep hills or towing, or when you need slow speed for manoeuvring or turning in confined spaces.

Top, or fourth, gear should not be used on heavy going.

To steer, move the handlebar firmly and steadily without snatching at it. If the compressor or air system is not working, additional effort will be required to steer, but the system will still function. Always make sure that the steering bar returns to the straight-ahead position after steering.

For steering in forward gears, use the steering handlebar in the same way as you would on a bicycle – to turn right pull the right-hand end and vice versa.

Steering in reverse is, in effect, the opposite. Push the right-hand end of the bar to swing the tail of the tank to the right and vice versa.

Steering in neutral involves moving the bar in the same way as for steering while driving. The tank will pivot in little more than its own length. Never do this on heavy ground (or where you are worried about damage to the surface you are on).

A note of caution about driving in confined spaces: make sure that you are guided by someone outside the tank who can see where it is in relation to its surroundings and where it needs to go. This person has to remain in the line of sight of the driver and use hand signals to indicate direction in forward, reverse and neutral turns. It is possible for this guidance to be given by a person in the commander's or loader/operator's positions, using intercom or radio microphone headsets. However, someone in the turret is less well placed to judge gaps and cannot see as much as a person on foot.

There are generally recognised hand signals for directing tracked vehicles. These need to be given clearly and decisively, with obvious, unambiguous movements of arms and hands and the person directing must make sure that the driver can see them. This person also needs to understand how to direct steering in reverse – in other words, that the directions refer to the way in which the rear of the tank will move. Neutral turns require a different signal and both driver and guide must agree this.

When crossing an obstacle, proceed slowly and steadily in first gear if crossing a large hole or trench. Cross the hole squarely, as doing so at an angle can force a track off.

Approach knife-edges, vertical walls, sleepers and similar low obstacles square-on and slowly. Tackle them gradually and steadily so that the tank rides over the point of balance smoothly. Take care not to steer or, if necessary, to use only very slight movements, when near to the point of balance. Use the brakes carefully; they are powerful and effective and should not be abused or used harshly. Engage a low gear when going down a hill to avoid overstraining the brakes. Never allow the tank to gain so much momentum on a downward slope that maximum braking effort is required to control it. If you do, excessive wear will result to the brake linings and indeed may burn them to the point of ineffectiveness. Also, take care when using the engine as a brake, that the rpm do not exceed 2,000.

If stopping for more than a brief period, switch off the engine, set the petrol control to 'off' and turn off the battery master switch. Replace the safety catch on the fire extinguisher bottle if the original system is in use. When you have finished running the tank and put it away:

■ Switch off the engine.
■ Set petrol control to 'off'.
■ Turn off battery master switch.
■ Replace safety catch on fire extinguisher.
■ Refill petrol tanks if desired.
■ Check water and oil levels.
■ Engage first gear and apply handbrake.
■ Check free travel of clutch pedal. If it is down to an inch or less, the clutch needs adjustment.
■ Examine petrol, oil and water connections for leaks and weeping.
■ Test by hand the temperature of brake drums, final drives suspension and bogies.
■ Inspect tracks and suspension for slackness and damage.
■ Make sure that the final drive spigot bolts and coupling bolts are tight.

Some 'dos' and 'don'ts'

■ Don't drive fast on frozen bumpy ground as the tank may get out of control. The Churchill's grip on icy surfaces is not good.
■ Don't run with slack tracks – they will not only affect steering, but also may come off.
■ Watch the rev counter when driving.
■ If you lose air pressure, steering and gear changes are still possible but they will be harder work.
■ Don't try to cross obstacles with petrol at a low level. Some fuel gets trapped when the tank is tilted leaving the suction pipes above the fuel.
■ Don't run the tank with the cooling system short of water. A dry system may lead to cracked castings and blown gaskets.
■ Don't ride on the clutch – it can lead to a slipping clutch and overheating.
■ Use a funnel for oil filling. Spilt oil is easy to mistake for a leak.
■ Open the water filler caps slowly after a run and protect your hands. The system operates under pressure and a jet of steam or boiling water may erupt.
■ After checking and topping up the oil level, particularly if the tank has been standing for a long time, run the engine for a few minutes and re-check the level.
■ Be careful with all hatches and make sure that they are securely fastened when open – they are easy to drop and can cause injury; there was a condition known to crews as 'Churchill Fingers', or lack of them, caused by dropping hatches.
■ Always run with the engine hatches shut, unless checking for leaks. The air-cooling system will not work properly with these hatches open, as the air will not be drawn through the air louvres and the radiators, but instead take the 'easier' route through the open hatches.
■ A build-up of water or other fluids in the engine compartment can (and should) be cleared by using the petrol dump valve in the fighting compartment. Be mindful of requirements for disposal of any engine fluids and the risk of contamination when disposing of any oil or other liquids from the tank.
■ Do not drive jerkily or use violent starts – these will impose undue strain on the transmission system.

FAR RIGHT The cap badge of part of the 107th Regiment RAC. The crew shown in the photographs on page 138–143 are from this regiment. *(Author's collection)*

■ Do not completely empty a petrol tank before switching to the other side. The considerable suction generated in an empty tank will overload the petrol pump.

■ Always use the correct brake fluid and oils.

■ When filling the radiators or washing the tank, make sure that water does not accumulate in the engine compartment, as it will be blown on to the steering brakes by the fan and make them ineffective until they dry out.

Fighting in the Churchill: the crew and their roles

A manual for the Churchill tank that talks of it just as a machine would be very incomplete. The tanks were only as good as the men who used them, and as the war continued their tactics were refined. With a skilled crew, the Churchill tank was very effective. However, mistakes cost lives, and the difference between a successful battle and a lethal failure would often come down to human error rather than bad design.

One tank and its crew

These remarkable photographs of the crew of a Churchill preparing before a battle give a sense of the crew as a close-knit group going about their duties. The crew of a tank depended on each other, and lived in extremely close proximity for long periods. Not only were they trained in their own particular tasks and able to carry out each other's roles, but they were also trained in first aid, and a very comprehensive

first-aid kit was carried in the tank, including morphine to administer to badly injured crew.

At times the crew of a damaged tank had to be self-sufficient while those around them fought on. Medical and mechanical assistance was not always readily available.

The tank in the photographs is called *Briton* and was a very early Mark VII (the turret lacks the 'bulges' either side of the gun, has an aerial for the WS38 in front of the gunner, rather than at the back of the tank, and the cupola is an early two-periscope one, with an early blade vane sight). The tank was serving with B Squadron, 107th RAC, part of 34 Tank Brigade. These photographs were taken by an Army photographer, Sergeant Hardy, on 17 July 1944 and are in the archives of the Imperial War Museum. The photographer gave a description of the occasion:

'The big attack which has opened in Normandy, where hundreds of tanks have been thrown into the battle, draws attention once again to the large part played by the tank in modern warfare. These huge heavily

RIGHT Lt Fothergill shaves using an improvised dressing table. *(IWM B7606)*

FAR RIGHT Tpr MacGuinness hangs washing across the front of a Mark III fitted with a 6-pounder. On the track guard a Thermos flask and plates can be seen. On the glacis plate can be seen one of the two drinking water containers that are stowed in the tank. *(IWM B7609)*

armed fighting vehicles may be the deciding factor in the last stages of the life and death struggle with Germany. … These pictures are of a crew waiting to go into the present battle somewhere in Normandy and show how they live whilst preparing for their zero hour, how they became welded into one family that eats, sleeps, works and fights together as a crew.'

The crew were:

Lieutenant John Alvin Fothergill, aged 28, from Bishop's Stortford – the commander.

Trooper E. ('Mack') MacGuinness, aged 29, from Liverpool – the turret gunner.

Corporal Stan Walmsley, aged 29, from Manchester, the loader and wireless operator.

Trooper Jimmy Swain, aged 28, from Preston in Lancashire – the driver.

Lance Corporal David Thomas, aged 28, from Liverpool, the front gunner/co-driver.

The photographs were taken towards the end of the fighting by the regiment on Hill 112. A number of them were printed in *Picture Post* on 12 August, and on 14 August, in the battles around the River Orne, *Briton* was holed through the front of the turret by an 88mm round. Lieutenant Fothergill was killed and other crew members injured. The tank was recovered and taken to the tank repair depot at Villers-Bocage. Lieutenant Fothergill is buried in Banneville-la-Campagne War Cemetery (grave reference XI. C. 20).

LEFT The start of the day. Dave and Jimmy (Tpr J. Swain) clean the gun. *(IWM B7615)*

BELOW Jimmy hammers out a track pin to fit an additional link. *(IWM B7617)*

ABOVE Jimmy tightens the track adjuster after fitting the new track link. (IWM B7621))

ABOVE RIGHT Jimmy, as driver, has to take care of the engine too. (IWM B7618)

RIGHT Cpl Stan Walmsley, the loader, takes 75mm ammunition from Lt Fothergill and stows it in the tank. The two kinds of ammunition visible are HE and AP. The round in Cpl Walmsley's hands is HE. Lt Fothergill is holding an AP round and above it, an HE one. Behind the loader's hatch there are two HE rounds and one AP between them. Note that the rounds do not have protective clips on their bases. (IWM B7619)

LEFT Lt Fothergill checks and arms the hand grenades. Above him the Bren gun with its circular magazine can be seen. The plaque on the stowage bin behind him says that the Bishops Stortford Urban District Council raised money for the tank as part of the Summer Savings Campaign. Bishop's Stortford was Lt Fothergill's home town. *(IWM B7622)*

BELOW A well-known picture of one of the Besa machine guns being cleaned. Note in the background a crew member fitting the aerial for the WS38 in the original position. 'Briton' was a very early Mark VII and did not have the fitting for a WS38 at the back of the turret. *(IWM B7624)*

RIGHT The Besa ammunition has been checked and is being re-packed into the liner, ready for use. Note the revolver on the crewman's belt. *(IWM B7627)*

FAR RIGHT Camouflage netting and branches are draped over the tank. *(IWM B7628)*

RIGHT The squadron leader, Maj Hugh Davies, briefs the troop leaders, with Lt Fothergill on the left. The troop leader nearest to the camera has his revolver tucked precariously into his belt. In the background is a Mark VII, but not *Briton*. Maj Davies won a Military Cross for his leadership of B Squadron when crossing the River Orne on 8 August. In that day's battle, nine of the regiment's tanks were put out of action including that of Davies. *(IWM B7629)*

ABOVE LEFT Lt Fothergill in turn briefs the other tank commanders in his troop. In the background is a Mark III with 6-pounder. *(IWM B7631)*

ABOVE One of the crew stands guard over the tank with a Sten gun during the evening and night. *(IWM B7633)*

LEFT L/Cpl David Thomas in the front gunner's seat, Lt Fotherghill in the commander's position and Cpl Walmsley is the loader/operator. Note that the front gunner has a different microphone from the commander and turret crew. The turret has netting around it to help attach branches used as camouflage. One can just make out the fact that the mid-section of track guard has been removed where it passes under the turret. *(IWM B7635)*

MAJOR GENERAL E.V.M. STRICKLAND CMG, DSO, OBE, CSTJ, MM, STAR OF JORDAN (1913–82)

Following distinguished and decorated service in France Strickland was promoted and posted to 25 Army Tank Brigade where he commanded HQ Squadron. It was in this role that he witnessed, experienced, trained with, and participated in, the development of the infantry tank. He took part in the introduction of the then new Churchill tank to the regiments of the brigade.

With 25 Army Tank Brigade in North Africa from February 1943 he took part in many of its actions. Most important of these was resisting the German attack at Hunt's Gap. If the holding of this position failed, Beja would have fallen and the First Army would have had to withdraw into Algeria. The outcome would have been incalculable: the war as a whole might have been lengthened by at least another year.

North Irish Horse (NIH), the senior regiment of 25 Tank Brigade, was rushed into the Hunt's Gap defence at the close of February, just in time to play its part in resisting the German

assault. By 3 March, one squadron leader had been seriously wounded and subsequently evacuated, and another squadron leader had been killed. North Irish Horse asked Brigade to release Strickland to them urgently. Strickland took command of 'A' Squadron NIH in Hunt's Gap where the German assault was pressing.

In early April, NIH took part in the Battle of the Peaks, among the hills to the east of Oued Zarga and north of Medjez-el-Bab. In that series of actions, 'Strick', as he became known, commanded NIH 'A' Squadron with

success, and this and the regiment's other achievements did much to confirm the fast-developing fame of NIH in infantry support. There, in that rugged upland country, NIH also established the reputation of the Churchill tank in the unprecedented use of tanks in intractable country.

From 23–26 April, Strick commanded 'A' Squadron of NIH in the memorable action at Longstop Hill, alongside 'B' and 'C' Squadrons and the exceptionally gallant infantry. Their determined use of Churchills in surmounting the prepared German defences on the northern slopes of the hill played its part in the achievement of a great tactical victory. With Longstop taken, the Medjerda Valley was accessible at last to the final advance on Tunis. In the capture of Tunis in early May, Strick's 'A' Squadron experienced two 'victory' passes through the city centre, taking up a position against a potential German counter-attack from the Carthage direction.

After the North African victory, the tank brigades withdrew to the Algerian coast, where they remained, unused, for nearly a year. In July 1943, promotion came for Strick when he went to 51 RTR as second-in-command. There he remained until January 1944.

In March 1944, Strick was promoted to lieutenant colonel and his first regimental command: 145 RAC. It was in this capacity that he gave the lectures on tank warfare. He took 145 RAC to Italy in April 1944 but, almost immediately, was ordered to return to command the NIH in the Hitler Line battle and breakout to Rome. In that memorable action, NIH played a key role in a striking victory, for which they have been remembered ever since. Strick returned to 145 RAC in June 1944, and commanded them in the Gothic Line battles.

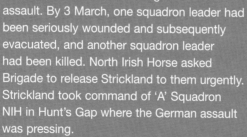

Tank tactics learned the hard way

I am fortunate to have been given access to the previously unpublished lecture given by Lieutenant Colonel E.V. Strickland MM, to 145 Regiment RAC on 24–26 April 1944 shortly before they joined the fighting in Italy. This fascinating document sets out the very experienced commanding officer's view of the lessons to be learned from the troops' experience in Churchill tank combat to date, taking account of fighting in North Africa as well as reports he had received from Italy. It provides a very important insight into the soldier's role in the tank, as well as the Churchill's capabilities and limitations.

The counter-attack in support of infantry

Our tanks were made to support infantry, and the fact that infantry require tank support is the reason for our existence. This will be our ordinary role at all times when the front becomes static. Our job will be to place ourselves in such positions that we can give the infantry immediate fire support against a German attack on the infantry positions, or to counter-attack quickly in order to restore any of our infantry positions that have been lost.

This will mean that a squadron is sited so as to cover all the likely enemy approaches to the infantry positions or placed behind the infantry covering their positions. Sometimes it may even be necessary to put one troop or two troops well in front of our infantry defence lines. Careful reconnaissance will be the answer to this role and then good siting of troops of tanks, well camouflaged, so that they cannot be easily dispersed by enemy shelling or bombing. Their positions must dominate the ground over which the enemy, either tanks or infantry or both, will have to move in order to attack our infantry. Alternative positions will be thought out carefully. It may well be that the tanks will have to stay out in their positions for many days and nights, so that we must learn to arrange fixed lines for our fire during the hours of darkness. [This meant giving gunners directions in which to fire 'blind', having made sure that the lines chosen would not cause friendly casualties.] The tremendous fire power of a squadron of Churchills at night will act as a deterrent to even the best German troops.

In Hunt's Gap, near Beja, a squadron of

Churchills remained in a position, known as the Loop, for nine days and nights and this position was approximately half a mile in front of our infantry forward defence lines. This squadron held the key to Beja and although Von Arnim tried hard to get through, he never succeeded. German patrols often worked through the squadron at night but once the alarm was given all the tank guns opened up on fixed lines and it had a really remarkable effect on the enemy. They were too frightened to counter-attack the tanks with sticky bombs or magnetic beehives.

The 'Set Piece' battle

This is our main role in offensive operations. Success will depend on first-class co-operation of all arms and that is the reason why I am insisting on your learning everything possible about infantry, artillery and sappers [Royal Engineers], their methods, capabilities and inabilities.

The object will always be to put the infantry on to the objective with the minimum of casualties and to remain there until they feel safe to release you.

Town and street fighting

This is not a type of fighting that we like, but neither is it nice for the infantry. We can help in street fighting and our job will be to provide the fire support for infantry as they assault houses and strong points. Experience has shown that we can do this well and it is obvious that the Churchill will be the best tank for this job.

ABOVE Infantry and tanks working together. This picture was taken as 5th Division advanced towards Lübeck on 2 May 1945. A pocket of resistance was encountered involving Spandau, small arms and mortar fire. Infantry of 2nd Battalion, The Wiltshire Regiment, worked closely with Churchill tanks of 6th Guards Tank Brigade to clear the enemy. (IWM BU4990)

ABOVE Churchill tanks of 6th Guards Tank Brigade and infantry of the 2nd Gordon Highlanders advance from Donsbrüggen into Kleve on 11 February 1945. This image perfectly captures the role of tanks and infantry working together in a built-up area. The tank in front is a Churchill Mark IV or VI with 75mm gun. The tank behind is a Mark IV armed with a 6-pounder. Note that the infantryman walking alongside the lead tank is carrying a WS38 Set. The lead tank has extra track links used as armour and its turret has been camouflaged. *(IWM B14611)*

With your guns, you will be able to do tremendous damage to houses if you use them properly. Remember that high-explosive ammunition will do little damage to a wall and that Besa will not penetrate bricks, so you must aim high-explosive ammunition and Besa into the windows and doors. Armour-piercing ammunition will make a hole in a wall for you if there is not one already.

Don't open up in streets [meaning to open the tanks' hatches and become visible and vulnerable] unless you are absolutely certain that you are in a safe area. The Canadians lost many tank commanders to snipers. Watch also for grenades and parachute bombs that may be dropped from tops of houses. Closed down you cannot be hurt.

Our motto must be that we may lose the odd tank or two to anti-tank guns but no other German trick will beat us.

The advance

Italy has given us the clearest picture of tanks in the advance. Usually this develops into an advance down a road and the old problem of the leading tank being knocked out at a bend in the road remains. Normally, the advance is led by reconnaissance elements but it has been found that this is too costly in reconnaissance personnel as the German mine or anti-tank gun not only destroys their small vehicle but also kills the whole crew. A tank only loses one track or bogie to a mine and even the notorious 88mm anti-tank gun will hurt or kill only one or two members of a tank crew. Therefore we must be prepared to lead advances.

LEFT L/Cpl Lodge of 278 Field Company, Royal Engineers, holds a German hollow charge *Hafthohlladung* 3.5 anti-tank magnetic 'Beehive' mine during Operation 'Epsom', 26 June 1944. Around its base are three magnets enabling the device to be attached to a tank (if it did not have anti-magnetic coating).The mine was set off by pulling a cord attached to the igniter and it was able to penetrate some 140mm of armour. In street fighting the Germans would wait until a tank had fired its main gun and rush out of buildings amid the dust to fix these mines to the side of the tank. *(IWM B6015)*

Close country fighting

Italy and to some extent S. France and Austria are covered with olive groves, vineyards and plantations. In the Italian campaign it has been found that the Sherman is not satisfactory for this type of country as it is too high and moreover its hull gun is not low enough. Your Churchill will be extremely useful in olive groves, etc. as it is low and the hull gun [the Besa] is literally 3 feet from the ground. Thus the hull gunner or co-driver will come into his own, as he will be able to fire under the low branches etc. It is difficult fighting this, and the advantage lies with the well-sited German tank and anti-tank gun, but with an alert hull gunner we will be able to overcome the difficulties. Moreover, our 6-pounder, using high-explosive ammunition, will be very effective where the Sherman 75mm high-explosive shell shows a tendency to explode on branches or twigs.

Some general remarks

Success in all these various roles of ours will depend upon the fighting efficiency of our tanks and crews. There is no room for self-complacency – there is always room for improvement. No tank must be lost without producing its effect on the enemy. Far too many Churchills and their crews were destroyed purely because time had not been spent in checking over the smallest details. In future, in this Regiment, everything must always work – revolver ports, pannier doors, engine covers, hatches, periscopes and telescopes. The cupola must be made to revolve very freely and must be kept in that state. On many occasions during the last campaign in Tunisia, Churchill tanks were really blind and only because their tank commanders had not taken the time or trouble to ensure that the cupolas and periscopes were in good working order. Remember, if any part of your tank does not work in action, then the answer is death for you and your crew and dismal failure for your troop, if not the whole Regiment.

Make certain that the turret flaps work smoothly and easily. One tank near Medjez El Bab got a direct hit from a Stuka. The crew were not even injured but the tank started to burn, and then it was found that the turret flaps would not open easily. They had not been greased. The answer was death from burning.

You have a petrol filler [the funnel to stop petrol spilling when poured] on your tank now. It is very important. Drivers, make certain that your tank is filled with petrol and oil very carefully and that engine oil and petrol are not allowed to accumulate under the engine [where they would form an inflammable sludge mixed together]. Many Churchills 'brewed' up in Tunisia when hit by anti-tank guns. It was found that the shot in some cases had penetrated the hull but not damaged the engine or fuel pipes, but that the friction of the high-velocity shot generated heat which ignited waste oil and petrol in the bottom of the tank. Don't let this happen to your

BELOW Churchill tanks of 25 Tank Brigade have moved up to support 1st Canadian Division in Italy on 17 May 1944. Here, maintenance is being carried out and ammunition is being passed into the turret of a Churchill of the Royal Tank Regiment shortly before the Hitler Line battle. The tank is a Churchill Mark IV with a 6-pounder gun and being passed in is an Armour Piercing, Capped, Ballistic Capped (APCBC) round. Behind the commander's cupola is an unusual-looking metal structure that appears to hinge forward and act as an overhead cover, but it does not look strong enough to stop a sniper's bullet. One member of the crew is fitting the 'B' Set aerial at the back of the turret while Sgt Bradstreet, from Bradford, and the driver, G. Stewart, deal with the ammunition. This was the first time Churchills had been used in Italy. *(IWM NA14974)*

Churchill, and remember that you can be hit by the 88mm many times and holed by them without suffering and without losing your battle. If you are idle, the first hit may destroy you.

The tendency is to say 'my lid never sticks' and 'my engine is clean' but unless you constantly check you will meet the same fate as the Churchills I have referred to.

Gunnery

Unfortunately, in the training of British tanks we have never put sufficient stress on gunnery, and yet it is the most important thing in our lives during war. The tank gunners, both turret and hull gunner, are the most important members of a crew.

Until recent months, British tank gunnery was not as good as that of the Germans. We were outgunned perhaps, but our guns were, and are, effective weapons if used properly. Our telescopes, while not having as great a magnification as those of the Germans, are satisfactory. The failure of our tank gunnery can be traced to the fact that we did not emphasise the importance of gunnery. Follow carefully then the following points that I am going to bring out.

We may have in Shermans in our organisation and we may have 75mm guns mounted in our Churchills. [This is a reference to the NA 75 conversion of some Churchills in North Africa which Lieutenant Colonel Strickland had seen taking place near to where he gave the lecture. See Chapter 2.] Churchills in England are being produced now with 75mm. The 6 Pounder is by far a better anti-tank gun than the present 75mm. At a test held near Beja in May 1943, representatives of 18 Army Group proved this. The front of the turret of the German PZKW VI (Tiger) is approx. 102mm thick. The 6 Pounder penetrated it at 300 yards – the 75mm could not penetrate at 10 yards. Both guns will penetrate the sides and rear of the Tiger and all types of German Mk. III and Mk. IV tanks at normal battle ranges.

Good gunnery depends on the cleanliness and careful preparation of guns. In future, whether we are static or in actual operations, guns will be T & A'd [T&A was an abbreviation for Test and Adjust] daily and tank commanders and gunners will do five to ten minutes' practice daily in giving fire orders and gun laying respectively.

On one occasion, in Central Tunisia, a squadron of Churchills formed up for battle. During the battle some of the guns would not fire, and others seized up after firing one round. All due to bad preparation and idleness. The gunner can never be satisfied that his 6 Pounder, Besa and ammunition are perfect. Check constantly.

On another occasion and near El Aroussa, 'C' Squadron of 142 Regiment R.A.C. completely destroyed a force of Germans, including seven Mk. III tanks. [In fact four were knocked out and three more disabled.] All the Churchills had fired many rounds, but it was found afterwards that most of the damage had been done by one good gunner who had always ensured the accuracy of his guns.

The essence of good gunnery is good indication of the target. This is the job of the tank commander, and unless he does it well he is letting down his crew. A bad tank commander is usually responsible if any of his crew are killed. He must give his gunner a good chance. Fire orders must be clear, vanes must be checked to ensure that they do line up the commander's periscope with the field of vision through the gunner's telescope.

The main difficulty in the gunnery is the judgement of distance. If sights are properly T & A'd, the high velocity of your 6 Pounder will ensure a hit even if there is an error of 200 yds. range up to 1,000 yds. Tank commanders are all able to estimate range up to 1,000 yds. with a maximum error of 100 yds. either way. Therefore, you must always hit with the first round at ranges of up to 1,000 yards.

The Germans in warfare put great emphasis on psychology. They say that British tanks cease firing once they have hit or think they have hit. This is true and we must make them change their ideas. In future, never stop firing at a target until you are absolutely sure that it is dead. Once you are on [target], go on pumping shots into it at a rapid rate. Remember that a hit does not mean the destruction of the tank. I remember examining a Churchill that had received 30 hits at ranges varying between 600 and 1,500 yds. Only 12 of those hits would have had some effect on the crew or the guns. The solid shot projectile just makes a hole. It is the same with the German tanks and now you are going to pump shots into them until they 'brew up'.

Don't be fooled. Once near Beja a Mk. VI (Tiger) fooled two Churchills. The Churchills hit first and penetrated. The German pretended he was done, and then suddenly came to life and destroyed both Churchills. Had they pumped shots into him, this would not have happened.

The main use of your Besas will be the domination with bullets of an enemy area or heavy shooting against a German counter-attack. The real way to do this is by spraying. It is useless pouring bullets into the parapet of a slit trench [this is the sloping ground just in front of the trench]. The ground must be thoroughly sprayed to keep the enemy down. Guide the gunner by the hose-piping use of tracer until he is on the target area, and then he can spray effectively. … Each tank will link up its sprayed area with the next one. In close country such as olive groves spray everything, even bushes that show no signs of life. An occasional 6 Pounder high-explosive shot adds to the effect on the enemy. Remember that spraying will give your fire support the added value of many ricochets. When spraying, don't fire in bursts but a complete belt at a time. When you are doing an aimed shot with the Besa, fire rapidly in short bursts of five or six rounds per burst. Check your aim the instant between bursts.

To counter attacks by Stukas and tank busting aircraft (usually ME 110 mounting 37mm quick-firing cannon) or fighter bombers … have your support troops of tanks on higher ground than the assaulting troops and when attacked by low flying aircraft, every tank will close down and all tanks will fire both Besas into the area of the aircraft. In this way the area above you will be filled with lead. In the Oued Zarga battles the North Irish Horse brought down an enemy aircraft in this way. … If we are sensible and close down quickly, no German aircraft can hurt us. A direct hit of a 500lb bomb did very little damage to a Churchill and some of the crew did not know it had hit them. Never use your Bren from the turret against aircraft. Don't foolishly expose yourselves.

Remember that when you have fired a few rounds from one position the enemy have discovered it, so move to a new position. This does not mean that you can only fire one round

from each position. NO, destroy your target as I have said before, and then move to a new area, even if it is only 100 yds. away. This will upset the German observation post, who may be planning to bring heavy-calibre shells on to you. …

Only heavy shell direct hits can hurt a Churchill (150 and 210mm), but it is very difficult to hit a tank, and if you camouflage your tank well and move about as I have suggested, they will never hit you.

In future, only use the much talked of hull-down position for firing. [Hull-down refers to a position where the tank is partially hidden behind an obstacle or in a shallow hole and only the turret is visible above the ground or obstacle.] The proper position for any tank is turret down, unless it is actually firing. German anti-tank gunners know the Churchill – a number were captured at Dieppe – they know the weakness of the turret mounting [the internal mantlet]. … But they can't hit the mounting if you are turret down and observing.

Seventy per cent of the Churchills hit by German anti-tank guns in Tunisia were holed in the gun mounting. The reason – hull-down positions. I have known Churchills to remain in hull-down positions for hours on end without firing. This gives the enemy time to bring up an 88mm anti-tank gun, or heavy tank. I remember once a Tiger Mk. VI taking up a hull-down position and remaining in it. It gave us the time

ABOVE A knocked-out German PzKpfw IV tank in a hull-down position, Normandy, 13 July 1944. The lecture notes that hull-down positions did not provide sufficient protection; a turret-down position referred to in the text involves the whole tank being behind cover. This tank had been destroyed by Allied aircraft.
(IWM B7056)

RIGHT Churchills of the North Irish Horse lay down indirect fire on enemy positions west of Mezzano in the Alfonsi sector, Italy, 21 March 1945. This photograph illustrates the use of NA75 Churchill tanks as artillery in indirect shooting. Note that the Sherman-derived external mantlet of the modified tank is very clearly visible. Littering the ground beside the tank are parts of the three-round packing cases for the 75mm rounds, as well as the inner cartons holding individual rounds and shell cases. Note the 75mm HE rounds stacked on the tracks ready to be passed to the turret crew. The tank nearest to the camera is from 2 Troop, A Squadron. Note that the pistol port on the turret is open, possibly for ventilation, and all but the front section of track guard is missing. (IWM NA23277)

(3 hours) to bring up a 17 Pounder anti-tank gun. It was just out of effective 6 Pounder range but the 17 Pounder got it.

High-explosive ammunition is effective in street fighting, but only if it is directed into windows, holes and doors. … Use armour-piercing ammunition to make holes in houses and always remember the German trick of hiding a tank in a house. If fire is coming from a house, and you know that your high-explosive ammunition and Besa is entering the door or window accurately, then it is a German tank or well prepared machine-gun post. Use armour-piercing ammunition if this happens.

Our Sherman and Churchills with 75mm guns will have a valuable observed shoot high-explosive role. There will be two main uses for them:

(i) 75mm tanks will be used to neutralise anti-tank guns in the battle that have not been located previously by reconnaissance or air photographs, and therefore not already accounted for in the support artillery fire programme. 75mm tanks will watch assaulting Churchills closely, and keep watch for anti-tank guns on the flanks. If they do not spot the gun themselves it will be indicated to them by the Churchills.

(ii) Immediately the objective is taken by infantry and [6-pounder] Churchills, the 75mm tanks will race to it and engage the enemy targets that are always exposed by the taking of an objective. These targets may be lorries, half-tracks, gun positions and enemy movement. … Don't imagine that your 75mm tanks are artillery pieces. … Our 75mm ammunition is limited. Only use it when you know that the artillery is too busy on other important targets or cannot do the job properly.

In static warfare, you may be called upon to do some 75mm shoots against enemy observation posts, etc. In offensive operations remember that your ammunition is limited, and that replenishment is not easy.

In action and whenever you are in an exposed position, keep your turret traversing slowly between three and nine o'clock. In this way the gunner will be viewing the country himself and he may spot something through his telescope. Secondly, a moving turret may just help to deflect the 88mm or 75mm shot.

The turret-down or hull-down position does not mean being behind a crest; sometimes the ground is flat and there are no features but there are always trees, bushes, hedges, walls,

haystacks, buildings and even knocked out tanks. The enemy will always find it hard to hit you, [even] if you [are] in a really weak place, if you are hidden. If solid cover is not at hand, find visual cover. Everything helps.

The Germans say that we, the British, use smoke under all conditions and that our smoke is good, but they have discovered that we are reluctant to move from behind it. In consequence German anti-tank gunners and machine-gun crews have been trained to fire all their weapons into the middle of all British smoke screens. … Make use of the few minutes of temporary safety that it gives you to move to a position of safety. In fact, one of the best ways of dealing with machine-gun posts and anti-tank guns is for the tanks to keep them smoked until the infantry are close enough to assault with the bayonet or throw grenades.

Sequence of action in battle

There is sometimes a tendency to forget the battle as soon as something difficult happens. For example, if a tank is hit, it is sometimes patent that troop leaders lose sequence with the result that the whole troop is lost. Owing to excitement, many tanks are lost trying to save the crews of a damaged one. In a battle, when

ABOVE A Churchill Mark IV with 6-pounder gun, probably of 51 RTR, uses bushes in the Italian countryside as natural cover, 20 July 1944. Note that the front right track guard is missing, while the one on the opposite side is pristine with the rubber still attached. The WS38 aerial base is unusual in that it is mounted on a bracket on the front left of the turret. Note also the very early basic blade vane sight. *(IWM TR2025)*

BELOW Heavily camouflaged Churchills of the North Irish Horse, 25 Tank Brigade, probably in the brigade concentration area at Arezzo in central Italy, 19 July 1944. *(IWM NA17041)*

a tank is hit by an anti-tank gun it is obvious that the squadron or Regiment is in the field of the enemy's anti-tank defensive fire so that the principle must be to speed up your attack rather than help the chap who is out of action. You must harden yourselves to casualties and remember that the operation in hand comes before succouring the wounded. I would like you to approach the problem in this way. Leave it to the troop concerned but learn from what has happened. This should be the sequence of thought and action when a tank is hit:

1. The squadron notes the fact that anti-tank guns have opened up but all troops less the one that has lost a tank and its support troop continue the battle.
2. Troop leader concerned uses his remaining two tanks to try and prevent the crippled tank from being hit again. This, either by destroying the anti-tank gun if located or by smoking of the cripple.
3. He must then ensure that his two tanks are in sound positions – both may have to scurry to cover.
4. Get on with the battle – this may mean directing infantry or artillery onto the anti-tank gun.
5. Report the casualty.

Don't waste precious time reporting a casualty until the moment is opportune. Remember that each crew is taught to fend for itself. It is better to capture the objective first and then return to the help of wounded comrades than to lose the battle altogether by trying to help a crippled tank on the start line. Remember also that we have a plan and an organisation for the collection of casualties.

RIGHT In February 1943, after two days of battle in front of Thala in Tunisia, British forces advanced several miles towards Kasserine. Here, a sapper of the 8 Field Squadron, Royal Engineers, lifts a Teller mine on the Thala–Kasserine road, which had been heavily mined by the retreating Germans, forcing armoured vehicles to take a cross-country route. Note the knife in the sapper's trouser pocket to help him dig gently around mines. This type of mine had a pressure-activated fuse on the top and held some 5.5kg of TNT, which meant it could blow the tracks off of a Churchill tank. The mines frequently had additional fuses on their sides and underneath to make them dangerous to defuse. Some could also be set off by a trip wire or pull-cord. *(IWM NA853)*

Wireless

The 19 Set is a very sound tank set and with proper attention it will ALWAYS WORK. … Remember that the success of the battle depends on perfect wireless communication throughout the Regiment – that the wireless is your lifeline. If anything happens to your tank, no one can help you if they cannot hear you.

Keep your batteries fully charged. The practice of running engines [to charge the batteries] at night when tanks are in the lines will cease. If a squadron has to stay out at night, then the Squadron Leader must put as many sets as possible off the air. The Germans realise that we are unpractised in night fighting and, as a result, they use the cover of darkness for most of their offensive work. We cannot afford to make noise at night as in night fighting one depends on hearing. If you run tank engines at night, no one for miles around will be able to hear anything else and the enemy will guide himself by your noise. Remember also that the 19 Set makes a loud hum. Therefore, at night, I want you to have just one set on and arrange for visual or oral alarm signals.

Mines

The mine is difficult to overcome. If we get to know mines the lifting of both S and Tellermines becomes quite a simple matter. The anti-tank mine is probably the finest weapon the Germans have against our tanks. In the big set-piece battle, where there is a known enemy minefield, sappers are produced to clear lanes for us but by far the majority of casualties due to mines

have been due to the unknown minefield. The Germans show a distinct tendency not to shell their own minefields and enemy mines have been picked up quite openly even in view of the enemy. The hull gunner or co-driver is best positioned to dismount and clear mines and all co-drivers will be so trained. … You must also learn something about British minefields. Learn where they are and if you are supporting infantry, particularly in static warfare, always take the trouble to ask your troop or squadron leader where the British mines are. If he doesn't know, insist that he finds out. Far too many British tanks have been knocked out on their own minefields. If you are anywhere near infantry, ask if they have put down any Hawkins grenades. [Grenade, Hand, Anti-tank, No 75, this was a small but effective grenade with just over a pound of explosive that could be laid on the ground as a small mine.] This grenade will blow the track and a couple of bogies off a Churchill.

Street fighting

The enemy usually lays mines at the crossroads and the crossroads are usually covered by a 75mm anti-tank gun. The mines are well buried and when a tank is stopped by the mines, it is knocked out by the anti-tank gun. Sometimes anti-tank guns are sited in cellars but usually they are outside the town with a good field of fire.

You can't use mine detectors so you have to take the risk. When you have to cross over crossroads, go absolutely flat out, even if this means reversing down the street you have just traversed to get the necessary 300 yds. run. The mines will probably shift your tracks, but the momentum will probably carry you into the cover of the buildings at the far side of the road and you will not be a sitting target for the anti-tank gun. Tanks alone cannot take a town or village but tanks, infantry and artillery can, so we shall always be supporting infantry in street fighting. Remember the German magnetic beehive mine [see page 146] and parachute bomb. Make sure when you plan to move your tanks that the infantry are going to watch them. In street fighting, we shall probably use a few tanks at a time. We lose formation and cannot operate as a troop. A lot of nonsense is talked about knocking houses down with tanks. The Sherman is reported to knock down houses very effectively and there is no

ABOVE A Humber scout car crew keeps watch for the enemy at a crossroads in newly-liberated Cormeilles, Upper Normandy, as a Churchill Mark IV or VI with a 75mm gun burns in the town square beside the war memorial, 26 August 1944. The scout car mounts twin Bren guns with 100-round drum magazines. Smoke can be seen emanating from the front hatches of the Churchill, and the 75mm gun breech appears to be open because smoke is also coming out of the barrel. *(IWM BU166)*

doubt that the Churchill could knock them down even more effectively but don't forget that most European houses have cellars and if you knock down a house, you might find yourself in a cellar with the house itself on top of you and unable to get out. Pillars supporting balconies can be struck a glancing blow with the back of the tank and will perhaps cause the house to collapse, but remember that piles of rubble will stop your tank. It is much more preferable to set fire to the house with high-explosive ammunition or smoke.

Feeding in the field

Latest reports from Italy indicate that this is a very grave problem particularly when laying out in support of infantry for long periods. Tanks are divorced from their A Echelon vehicles and one case is reported where men did not get their food for six days. In future, 6 Pounder ammunition boxes will be used as food boxes. There will be two boxes for each tank with tank and troop numbers painted on them and a reserve in the A Echelon. The boxes will be filled at night and sufficient boxes for a

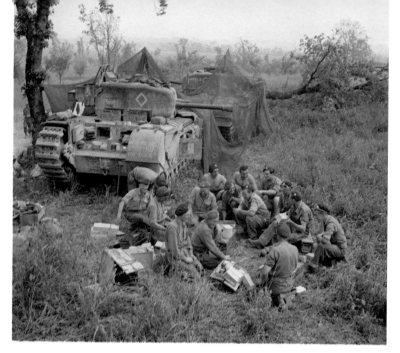

ABOVE Churchill tank crews of the HQ Troop, 51 RTR, share rations near their camouflaged vehicles before going into action in Italy in support of 1st Canadian Division, 17 May 1944. In the centre of the group, 6-pounder ammunition boxes are being used to store the food, as recommended by Col Strickland. The tank nearest to the camera is a Mark IV with a short-barrel 6-pounder gun. Another Mark IV, T68933R, can be seen in the background. These tanks, in common with the one shown on page 147, have an improvised metal structure behind the cupola. The legend 'Not to be stowed on deck' has been applied to the front left track guard of the Churchill in the foreground. This warning came about because two early Mark II Churchills (T30999 and T31000) were sent to Egypt in 1941 for trials and travelled by ship on, rather than below, deck. Although covered with tarpaulins, they arrived in what was described as a 'deplorable condition' because sea water and salt air had caused extensive rust and the destruction of alloy castings. *(IWM NA14976)*

complete squadron can be carried in one jeep. They contain 24 hrs. rations and there is also room for E.F.I. issues. The S.Q.M.S. [Squadron Quarter Master Sergeant] will not return until every box has been delivered to its tank. The tank crew will hand over their empty box for the full one. This system must always be used.

In battle, it is sometimes impossible to get out of the tank, but it is quite safe to cook in the co-driver's seat.

Evacuation of tanks in battle
First consider your tank as a fighting machine and what you have to do with it to complete your task. You have the right to save your lives if your tank is rendered useless. If your tank is on fire, you cannot hope to carry out any sort of

evacuation drill [which was a set of procedures to make the tank and its weapons unusable by the enemy and also to secure the removal of radio codes] and you should get out [of] your tank as quickly as you can. Try to take a weapon with you, but don't worry about anything else. I don't want you to risk your lives trying to carry out an evacuation drill when your tank is on fire. If your tank is knocked out and is not on fire, then the normal evacuation drill should be carried out. If it seems likely that your tank will fall into enemy hands, try to destroy it completely. Set it on fire, take out the striker mechanism [the firing pin element of the main gun], Besa breech block, remove maps and coder [the Slidex codes for use in radio transmissions], throw a grenade into the wireless set.

Shelling and mortaring
You will get shelled and you will get mortared. Always report back. Sometimes British guns have shelled British tanks and British guns will go on shelling British tanks unless the tanks report it and get it stopped. Even if you get a curt reply to information you have sent back, keep on reporting all you see in your area.

German tanks
I think it could be said quite fairly that the Germans have, to a good extent, used their tanks more cleverly in some ways than we have. We do not camouflage our tanks properly. In future, the moment you become stationary, even if only for ten minutes, you will camouflage your tank. You must try to make your tank merge into the background. Materials used for camouflage should be fastened to the tank. If you are in scrub try to find a position where your camouflage will not be higher than the surrounding scrub. Don't forget to camouflage the barrel of the gun, but always ensure that the camouflage does not interfere with your sights. A well camouflaged tank can be moved very, very slowly and yet remain undetected.

Shooting at German tanks
We have been taught, quite rightly, to shoot at what we call 'within the battle range'. If you are in a good position and German tanks are moving towards you, hold your fire until they are within nice range and you know that you can

smash them. If they stand off at 2,000–2,500 yds. and shell your positions with high-explosive ammunition and don't show any tendency to come forward, I want you to fire at them. This also applies to enemy tanks moving across your front. They may be doing a reconnaissance or feeling their way back for an attack, and you might frighten them off. There will be occasions, however, when you are hiding for a special purpose and in such cases, you will hold your fire and not give your positions away.

Tactics

Within the troop, tanks must always be supported, whatever type of battle is being fought. There must be a drill within the troop to ensure all-round observation. When given a task, always ask 'who is going to support me?'

The following points are reiterated for emphasis:

1. *We exist to support infantry and we must never say 'no' when asked to do something by the infantry. If the infantry's suggestion is impracticable then you will find an alternative method of doing what they want.*
2. *Plan for all-round observation within the troop.*
3. *Arrange for infantry to indicate targets to you by using Bren or rifle tracer.*
4. *Never get angry with the infantry. If the infantry don't turn up at the start line at the proper time, remember that things which won't hurt you will hurt the infantry, and perhaps stop them.*
5. *Movement of tanks – you will either be stationary or going flat out. The faster you go the harder it is for the anti-tank gun to hit you. Turret down, unless you are actually shooting. Ensure that someone is supporting you and then flat out for the next position.*
6. *Close down or open – depends on situation, use your common sense. If there is any question of you being overlooked by high ground or buildings, close down.*
7. *Lying out – we may sometimes have to lie out in support of infantry for many days and nights. Don't stay out unless you make a definite fire plan. It may be essential that some of the tanks switch off their wireless transmission sets to conserve batteries for the next day's operation, and so there must be a proper alarm system. There must*

always be one man awake and he should be armed with a tracer machine-gun and have his head out of the turret.

8. *Defence against gas and flame throwers – make sure your respirator [gas mask] is always in good order and always with you. The German tank flame thrower is not very effective against tanks. It is carried on Mk. IIIs and has to get within 75 yds., and a German Mk. III at 75 yds. is the tank commander's dream of a perfect target.*
9. *Avoid wadis [an Arabic term with which the troops would have been familiar in North Africa. It means a small depression, gully or dry riverbed]; they are usually mined and you will find yourself ditched and trapped. Remember that warfare must be flexible. Don't religiously keep to your course-line if the going is bad. Even if you have to move 500 yds. off it, do so. The Churchill will climb better than any other tank; use the high ground to jockey the enemy out of his positions. Longstop Hill was captured and the complete garrison surrendered when one Churchill reached the summit. By climbing you will outwit the enemy anti-tank guns.*

The old expression 'good tank ground' must be forgotten. The best going for tanks is obviously covered by anti-tank guns and heavily mined. Avoid the obvious approach. It may take longer to use bad going but you will attain your object.

Field, 24 April 1944, EVS/FM

ABOVE Churchill tank crews of 31st Tank Brigade with their extensively camouflaged vehicles after Operation 'Epsom', 13 July 1944. *(IWM B7078)*

How vulnerable was the Churchill?

One of the most important questions for the crews was: how vulnerable was the Churchill to enemy fire? In addition to the lecture cited above, there are surviving studies and briefings on this very important topic, and they help to separate myths from fact.

The 8 December 1943 'Report on the Analysis of Battle Damage to Churchills' by the Armour Section, School of Tank Technology, from the North African campaign, provides some very interesting statistics about the effect of German guns on Churchill tanks, all of which were earlier models than the Mark VII. Some 41 tanks that had been hit were studied for the report. Of these, 20 were damaged by mines. These 20 mostly also received further damage from gunfire, as mines were often deployed just to slow tanks or temporarily disable them so that they could be an easier target for guns.

The ratio of the effective to the ineffective hits for the three principal German guns against all tanks appeared to be as follows (doubtful rounds excluded):

Gun	Effective	Ineffective	% of rounds effective
88mm	13	7	65%
75mm	8	8	50%
50mm	2	5	29%

These figures, though they must be accepted with considerable reserve, are somewhat encouraging in that they show that a Churchill hit by the 8.8cm. is not necessarily finished. Therefore the armour is not completely outclassed by the present projectiles. There is some likelihood that a valuable increase in protection could be obtained by the addition of an inch or even half an inch of armour.

The Mark III and IV were given appliqué armour as a result of just this point.

The following very tentative conclusions have been drawn:

(a) The [German] 5.0cm. gun is obsolescent. The 7.5cm. and 8.8cm. were the principal calibres used in North Africa.

(b) The front of the turret and the side of the hull were hit most frequently. Each of these aspects receives about one third of the total hits obtained.

(c) About one in three of the hits on the front of the turret, and about two in three of the hits on the sides perforate.

BELOW Here is a view inside the tank where the *Panzerfaust* round hit, knocking off a disc of metal measuring some 6in in diameter. The explosive jet passed across the tank and struck the pannier door on the far side. To the left is the handle of the pistol port; to the right is the head of the bolt holding on the external armour.

RIGHT Just above the number '5' is the damage caused by a German *Panzerfaust* attack on the side of a Churchill, just in front of the pannier door. The following pictures show just how much damage was done on the inside of the tank. *(All photos on this page Author's collection)*

RIGHT This is the effect of the jet on the pannier door on the other side of the tank. The crater measures 3in x 6in, and is ½in deep.

(d) The difference in the armour basis of these two aspects is perhaps 18mm. This comparatively small increase in effective thickness seems to be responsible for halving the number of effective hits.

(e) The Churchill, though far from being immune from the 7.5cm. and 8.8cm. projectiles used in North Africa is not hopelessly out-classed. The addition of ½ to ¾ inch of armour all round would not make it immune, but would probably increase its resistance substantially.

The turret front and sides of the Sherman appear to be weaker respectively than those of the Churchill. If the report refers to typical battle damage, the prospects of the Sherman are not particularly promising.

One other weapon that was much feared by the tank crews was the *Panzerfaust* – a hand-held, shoulder-launched anti-tank 'bazooka'. This was frequently used by German patrols and could be very damaging in close-quarter fighting. These weapons used a shaped charge to penetrate the armour of a tank, and the blast could do significant damage inside the tank. One report of the effect of a *Panzerfaust* 100 fired at the side armour of a square-door Churchill refers to the blast having penetrated the near-side armour and caused a 6in-diameter disc to be punched out of the inner skin, and then formed a crater in the pannier door on the other side some 6in by 3in, and ½in deep.

ABOVE LEFT Churchill Mark III, T 172220, from 3 Troop, A Squadron, 9 RTR (denoted by the '992' with a diagonal stripe), fitted with appliqué armour (see the turret either side of the gun), has received a hit above the mantlet and low on the front. This must have taken place in north-west Europe. *(Tank Museum)*

ABOVE This view of Mark III, T172220, shows the shot that hit the turret and evidently penetrated it, peeling back the top armour. *(Tank Museum)*

LEFT A badly damaged Churchill Mark IV or VI. *(Tank Museum)*

FAR LEFT Another Mark III, holed just below the mantlet. The gun barrel has been shattered by a shot. *(Tank Museum)*

LEFT This Churchill Mark III, T69005R, is possibly in Tunisia and has received multiple hits on the turret and front. It appears to be in a 'graveyard' for tanks, judging by the background. *(Author's collection)*

Bibliography

Anderson Jr, Richard C., *Cracking Hitler's Atlantic Wall, The 1st Assault Brigade Royal Engineers on D-Day* (USA, Stackpole, 2009)

Beale, Peter, *Tank Tracks, 9th Battalion Royal Tank Regiment at War 1940–45* (Sutton, 1995)

Chamberlain, Peter and Ellis, Chris, *The Churchill Tank* (Arms & Armour Press, 1971)

Dyson, Stephen, *Tank Twins, East End Brothers in Arms 1943–45* (Pen & Sword, 1993)

Eastwood, Stuart, *Lions of England. A pictorial history of the King's Own Royal Regiment (Lancaster) 1680–1980* (Silver Link Publishing, 1991)

Erskine, David, *The Scots Guards 1919–1955* (1956, repr. Naval & Military Press, 2006)

Fletcher, David, *Mr. Churchill's Tank, The British Infantry Tank Mark IV* (Schiffer, 1999)

Forbes, Patrick, *6th Guards Tank Brigade, The Story of Guardsmen in Churchill Tanks* (Sampson Low, Marston & Co. Ltd, 1946)

Fortin, Ludovic, *British Tanks in Normandy* (Histoire et Collections, 2005)

Gudgin, Peter, *With Churchills to War, 48th Battalion Royal Tank Regiment at War 1939–45* (Sutton, 1996)

Henry Jr, Hugh G. and Pallud, Jean Paul, *Dieppe through the Lens of the German War Photographer* (After the Battle, 1993)

Hunt, Donald F., *To the Green Fields beyond* (Pentland Press, 1993)

Leakey, Rea, with Forty, George, *Leakey's Luck, A Tank Commander With Nine Lives* (Sutton, 1999)

Moczulski, Leszec, *Churchill Vol. 1* (Poland, AJ Press, 2011)

Smith, J.G., *In at the Finish, North-West Europe 1944–45* (Minerva Press, 1995)

Wilson, Andrew, *Flame Thrower* (Corgi, 1973)

Index